W9-BEE-108

POINT PATTERN ANALYSIS

BARRY N. BOOTS
Wilfrid Laurier University
Ontario, Canada

ARTHUR GETIS
University of Illinois
Urbana-Champaign

SAGE PUBLICATIONS
The Publishers of Professional Social Science
Newbury Park Beverly Hills London New Delhi

Printed in the United States of America

For information address:

SAGE Publications, Inc.
2111 West Hillcrest Drive
Newbury Park, California 91320

SAGE Publications Inc.
275 South Beverly Drive
Beverly Hills
California 90212

SAGE Publications Ltd.
28 Banner Street
London EC1Y 8QE
England

SAGE PUBLICATIONS India Pvt. Ltd.
M-32 Market
Greater Kailash I
New Delhi 110 048 India

International Standard Book Number 0-8039-2245-0
International Standard Book Number 0-8039-2588-3 (pbk.)

Library of Congress Catalog Card No. 88-60304

FIRST PRINTING

When citing a university paper, please use the proper form. Remember to cite the correct Sage University Paper series title and include the paper number. One of the following formats can be adapted (depending on the style manual used):

(1) IVERSEN, GUDMUND R. and NORPOTH, HELMUT (1976) "Analysis of Variance." Sage University Paper series on Quantitative Applications in the Social Sciences, 07-001. Beverly Hills: Sage Pubns.

OR

(2) Iversen, Gudmund R. and Norpoth, Helmut. 1976. *Analysis of Variance.* Sage University Paper series on Quantitative Applications in the Social Sciences, series no. 07-001. Beverly Hills: Sage Pubns.

CONTENTS

ACKNOWLEDGMENTS

Preparation of this work was supported by a WLU Instructional Development Grant and National Science Foundation Grant SES 82-10598. Some of the data were provided by William Hayden, Russell Muncaster, and Richard Turkheim. Grant Thrall, John Odland, and several anonymous referees provided us with helpful critiques of earlier drafts of this manuscript. Pam Schaus and Pam Carnochan drew the figures and checked some of the data and Lillian Peirce and Shirley Gates typed the original manuscript.

SERIES EDITOR'S INTRODUCTION

Maps are spatial representations of information, and when that information can be characterized by location on a map it becomes a spatial pattern. Spatial patterns, like other forms of information, can be analyzed statistically. This book is an introduction to the statistical analysis of such spatial or point patterns.

Many phenomena can be represented by points on a map: towns, stores and centers for shopping, industrial locations, parks, archaeological sites, plant and animal species, the home site of a person with a possible environmentally related disease, and so on.

The authors introduce the reader to the general analysis of the location of points on maps. Map patterns are assumed as having been created by one or more spatial processes in the human or physical world. Often the causal forces are known, but more frequently the researcher is seeking to identify the causal forces. The analysis of the spatial pattern of the phenomena under study can be a precursor for revealing the underlying causal relationships. Guidelines are provided here for the exploration of spatial patterns for situations where the causal processes can be either known or unknown.

The authors detail in a step-by-step manner the methods required for the analysis of spatial patterns. These methods generally evaluate the *dispersion* and the *arrangement* of characteristics of the phenomena under study. Dispersion measurements focus upon the density of the points, whereas arrangement measurements focus upon the relationship of the points to one another.

Professors Barry Boots and Arthur Getis are the leading authorities in point pattern analysis, and together translate the complex research on spatial point patterns for the reader who is new to the field. They assume that the reader does not possess mathematical skills more than that normally expected of first-year college undergraduates. By pitching the discussion at an introductory level the reader is freed from unnecessary mathematical derivations and statistical arguments. The emphasis is upon applications, so each method for the analysis of point patterns is accompanied by a clear informative example explained in a step-by-step manner. Students and researchers in such diverse fields as city planning, forestry, geology, archaeology, epidemiology and environmental health, and human and physical geography will benefit from this introduction.

—*Grant Ian Thrall*
Series Editor

POINT PATTERN ANALYSIS

BARRY N. BOOTS
Wilfrid Laurier University
Ontario, Canada

ARTHUR GETIS
University of Illinois, Urbana-Champaign

1. INTRODUCTION

A common concern that runs through the diverse branches of geography is the examination of the spatial occurrence of a particular phenomenon. For example, we may be interested in the location of towns in a state, industrial plants in a city, sinkholes in a karstic region, or polluters along a body of water. In each case it is possible to display the data in the form of a map. Through cartography, objects in the three-dimensional real world are displayed as symbols in a two-dimensional plane. Usually these symbols are of three basic geometric forms: points, lines, and areas.

One of the most common types of map produced using such symbols is one in which the occurrences of the phenomenon studied are represented as points. We shall call these point pattern maps. Although the real-world objects themselves are not points, such a representation is possible because the physical sizes of the objects are very small relative to both the distances between them and the extent of the area in which they occur.

Geographers examine point pattern maps for a variety of reasons. A major reason is their belief that such maps represent one source of evidence that may be helpful in learning more about the phenomenon represented and the processes responsible for generating it. Sometimes our ideas concerning a phenomenon are sufficiently developed that we may be able to build an explanatory model of it. Quite often hypotheses concerning the locational behavior of the phenomenon can be derived from such models. For instance, central place theory suggests that settlements should be regularly distributed over a region. Support for such a hypothesis, and thus the model from which it is derived, can be obtained from analysis of point pattern maps showing the distribution of central places.

Similarly, economic considerations contained within theories of urban rent suggest that individual occurrences of some activities will repel each other, thus dispersing the activity, whereas individual occurrences of some other activities may attract each other, thus producing spatial aggregations. Some types of retailing such as shopping malls are examples of the former activity; some types of industrial and office activities are examples of the latter. Again, such locational expectations can be tested by examination of point pattern maps.

Another example comes from the study of the diffusion of information. Various theories imply that new information spreads according to principles relating to the proximity of potential communicants and their susceptibility to receiving new information. These notions can be tested by analyzing point pattern maps showing different periods in time and different environmental conditions.

Even when our knowledge of a phenomenon is very rudimentary, information gained from the analysis of point pattern maps may enable us to acquire some initial insights into the phenomenon. For example, the discovery that drumlins are often spaced differently on the margins of drumlin fields than they are at the center of such fields may lead us to investigate the possibility of different forces operating at those locations or the possibility of the same forces operating but with different relative intensities. Similarly, the knowledge that incidences of cases of a little understood disease are widely dispersed over a region might lead us to consider that it was not spread by contagion.

In this book it is our purpose to introduce a sample of the procedures that have been used in geography to analyze point patterns. The presentation will stress those techniques and their derivatives that have been most extensively used in geography. Our objectives are to enable readers to evaluate the appropriateness of existing applications of point pattern analysis and to provide sufficient background for the readers to pursue examples of their own. In terms of the techniques presented, our presentation is selective and emphasizes typical problems that the reader is most likely to encounter in practice. Because of this emphasis, when formulae are used neither their derivations nor their proofs are presented. Those readers who are interested in such issues or who wish to pursue more exhaustive and sophisticated treatments of point pattern techniques are referred to the recent reviews by Cliff and Ord (1981: chap. 4), Cormack (1979), Diggle (1979a, 1983), Getis and Boots (1978), Ripley (1981) and Upton and Fingleton (1985). The literature in the field of spatial analysis is vast. For example, Upton and Fingleton cite about 500 books and articles.

In the remainder of this chapter we first review the different situations in which point pattern analysis has been employed and then examine the fundamental characteristics of point patterns and the processes that generate them. This material represents the stimulus for the development of this book. Our goal is briefly to present in as simplified form as is reasonable the main ideas and techniques of point pattern analysis.

1.1 Overview of the Development and Application of Techniques of Point Pattern Analysis

The origins of the techniques currently used in the statistical analysis of point patterns arose over 50 years ago in plant ecology. These early studies, which are now primarily of historical interest, are reviewed by Greig-Smith (1964: chap. 3). With the exception of a study in geography involving the location of villages in two areas of the Tonami Plain of Japan (Matui, 1932), most of the work on statistical point pattern analysis over the next 25 years continued to be in ecology, although during this time such work was extended to cover animals as well as plants. Plant and animal ecologists have used such techniques to explore both the spatial distribution of individual species and the interrelationships of two or more species. Their overall aim was to identify factors of the individuals and their environment that influence such patterns. In general, these techniques are most appropriate for subjects that have a fixed location, such as individual members of a plant or tree species and conspicuous and relatively immobile animals or features associated with them such as nests, food caches, or display sites. Earlier studies involving subjects as diverse as grasshoppers, frogs, snails, and beetles are reviewed in Southwood (1966: 39-40; 1978: 47-48). Examples of more recent studies include those of granary trees of acorn woodpeckers (Roberts 1979; Burgess et al. 1982; Mumme et al. 1983; Burgess 1983; Brewer and McCann 1985), pits made by ant-lion larvae in both experimental and field situations (McClure 1976; Simberloff et al. 1978; Simberloff 1979), caddisfly populations (Lamberti and Resh 1983), ant nests (Harrison and Gentry 1981; Levings and Franks 1982), and feeding cells of cicadas in soil (White et al. 1979).

In the early 1960s, as geographers entered a phase in the development of the discipline now referred to as "the quantitative revolution," the techniques developed by ecologists were introduced into geography to refine and substantiate previous qualitative descriptions, particularly of settlement patterns. Initially, most studies focused on examining the extent to which characteristics of settlement location predicted by central place theory (King 1985) could be identified in real-world situations (Dacey 1960, 1962; King 1961, 1962; Birch 1967). Soon after, geographers recognized the potential of such techniques in testing hypotheses concerning the processes responsible for the patterns they described. Under the leadership of Michael Dacey they began to develop and extend models of their own that produced alternative patterns to those proposed by central place theory. In particular, models leading to clustered patterns of settlements were emphasized (Dacey 1963 through 1973b; Dacey and Tung 1962; Harvey 1966, 1968a, 1968b; Hudson 1969, 1971; Getis 1974).

Simultaneously, point pattern techniques were extended to analyses of phenomena other than settlement patterns, in particular, retail establishments. Initially, they were used to identify patterns in the distribution of urban retailers (Artle 1959; Clark 1969; Rogers 1965, 1969a, 1969b, 1974) and changes in these patterns over time (Getis 1964; Lee 1974; Shaw 1978). The results of such analyses have also been used to rank retail functions and examine temporal changes in the rankings (Artle 1965; Sherwood 1970; Sibley 1972). More recent work has concentrated on exploring the relationship of spatial patterns of stores with other aspects of the retailing environment such as the distribution of customers (Guy 1976) and city size characteristics (Sibley 1975).

Geographers and others have also used point pattern analysis to examine the spatial characteristics of a number of physical features of the landscape including drumlins (Smalley and Unwin 1968; Trenhaile 1971, 1975; Hill 1973; Gravenor 1974; King 1974; Muller 1974; Jauhiainen 1975; Rose and Letzer 1975), cirques (Sugden 1969; Robinson et al. 1971; Unwin 1973), volcanic craters (Tinkler 1971), sinkholes in karstic regions (Williams 1972a, 1972b; McConnell and Horn 1972; Day 1976), tors (Bardsley 1978; Sneyd 1982), inselbergs (Faniran 1974), river basin outlets (Morgan 1970), and junctions in river channel networks (Dacey and Krumbein 1971; Oeppen and Ongley 1975).

It was primarily from geography that point pattern analysis was introduced to archaeology and anthropology. Here such techniques have been used in three circumstances: to study artifact distributions within a site, to study artifact distributions over an area, and to study the distribution of sites. Examples of each type of application are reviewed in Hodder and Orton (1976). Interest in point pattern analysis in archaeology has been such that recent contributions to theory have been made (for example, Pinder et al. 1979; McNutt 1981; Stark and Young 1981) as well as continuing empirical applications (for example, Adams and Jones 1981; Carmack and Weeks 1981; Ward 1983; Weeks 1983).

There have also been less frequent applications of point pattern analysis in other disciplines such as astronomy, where it has been used to study the distribution of galaxies (Neyman and Scott 1952, 1958; Neyman et al. 1956; Peebles 1974), and materials science, where it has been used to study the distribution of particles in metals (Werlefors et al. 1979; Wray et al. 1983).

Finally, in the last few years, a number of statisticians have developed more complex and wide-ranging techniques of point pattern analysis. Foremost among this group are Ripley (1981) and Diggle (1983).

1.2 Some Fundamental Properties of Point Pattern Maps

In general, a point pattern map contains two major types of components: the points representing the objects being studied and the geographical area in which they are located. We shall refer to these components as the *point pattern* and the *study area,* respectively. Before we discuss techniques for analyzing point pattern maps it is necessary to become familiar with some basic properties of both point patterns and study areas. Although some of these properties are quite obvious, they are important because they may influence the selection of an appropriate analytical technique as well as particular decisions that relate to making individual techniques operational.

One of the most obvious properties of a point pattern is its *size.* This is simply the number of points, N, in the pattern. The study area may be represented by features of various dimensions. For example, if we are examining the locations of stops along a transit line, clothing stores along a downtown street, or effluent discharge points along a river, it would be appropriate to represent the study area as a line (one-dimensional). If we are concerned with the locations of settlements in a region, past residences of recent movers in an urban area, or sinkholes in karstic topography, the study area will be two-dimensional. Similarly, in some geomorphological or meteorological instances, a three-dimensional study area might be more suitable. We shall refer to this aspect as the *dimension* of the study area. In this book we limit our attention to study areas of one or two dimensions. We use the symbol W to denote the length of a one-dimensional study area and the symbols A and B to denote the size and perimeter length, respectively, of a two-dimensional study area.

Two-dimensional study areas may be bounded in a variety of ways. We shall refer to the figure enclosed by the boundary as the *shape* of the study area. Thus study areas of one dimension do not have this property. For our purposes, the most important aspect of the shape is whether it is regular or irregular. Examples of such boundaries are shown in Figure 1.1.

Further, the study area may represent a real geographic unit, such as a state, a city, a trade area, a river, or simply be imposed on reality by ourselves or others. Study areas of the former type will be said to have a *real boundary,* whereas those of the latter type will be said to possess an *arbitrary boundary.* Usually—but not always—real bounded study areas are irregular in shape. For convenience those with arbitrary boundaries are more often regular.

Finally, we may study the location of the points in the pattern with respect to the study area or with respect to each other. In the former case we are examining the *dispersion* of points, whereas in the latter case we are studying the *arrangement* of the points. As we shall see, in many cases these two properties are highly correlated. However, since arrangement refers to properties of the relative locations of points, its study can be useful when the real boundary of the study area is unknown or difficult to define (for example, the limits of a particular soil or vegetation type or the margins of a waterbody) or where we do not wish to impose an arbitrary boundary.

1.3 Fundamental Types of Point Patterns and Fundamental Processes for Generating Point Patterns

We have suggested that the main reason we examine point pattern maps is to attempt to learn something of the processes that generated the pattern. When analyzing point patterns, geographers have most often made use of a scheme that involves establishing a theoretical pattern with respect to which other patterns are identified. The theoretical pattern chosen is one that results from the operation of what is formally called a *homogeneous planar Poisson point process.* In this process points are generated in a study area subject to two conditions:

(1) each location in the study area has an equal chance of receiving a point (uniformity); and
(2) the selection of a location for a point in no way influences the selection of locations for any other points (independence).

Alternatively, these conditions may be interpreted, respectively, as implying that the study area is completely homogeneous in all regards

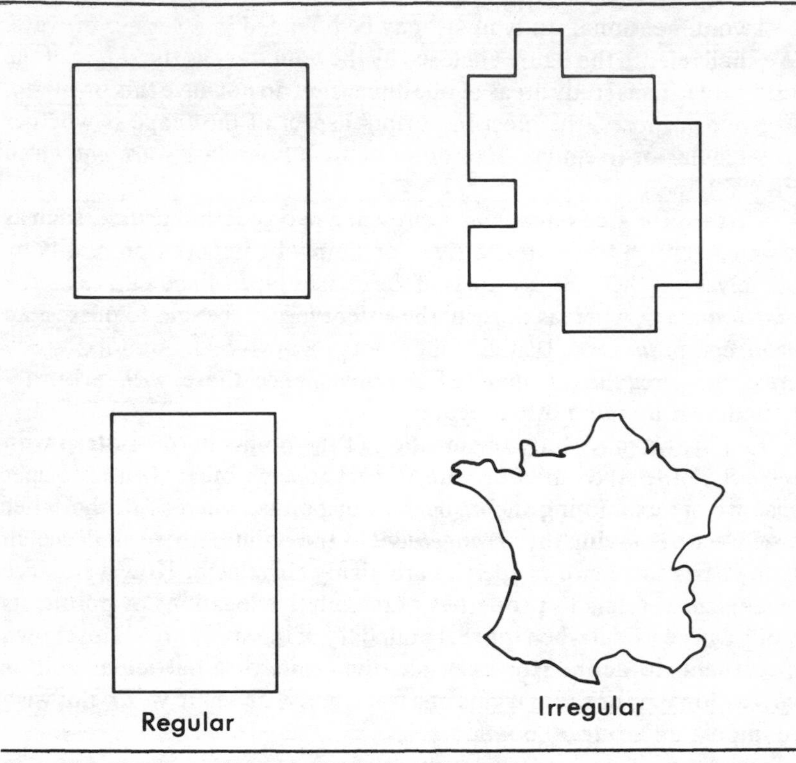

Figure 1.1 Regular and Irregular Study Area Boundaries

and that there is no interaction between the points. Hence the pattern that results from such a process can be considered one that would occur by chance in a completely undifferentiated environment. We will call such a pattern complete spatial randomness (CSR) after Diggle (1983). An example of CSR appears in Figure 1.2a. In view of the conditions involved in the generation of such a pattern, it is unlikely that true CSR occurs in any real-world situation. However, the processes acting in the real world are many and diverse and when no strongly dominant ones prevail, the net effect may be to produce a pattern that has the appearance of CSR, that is, the end product of the processes is to produce a pattern that is indistinguishable from CSR.

Our interest in CSR is primarily for its role as an idealized standard. The two conditions mentioned earlier that are assumed to exist when CSR results provide a simple model that can be useful in many circumstances. For instance, if we know little about the processes responsible for a particular pattern, we can begin by testing the hypothesis that the

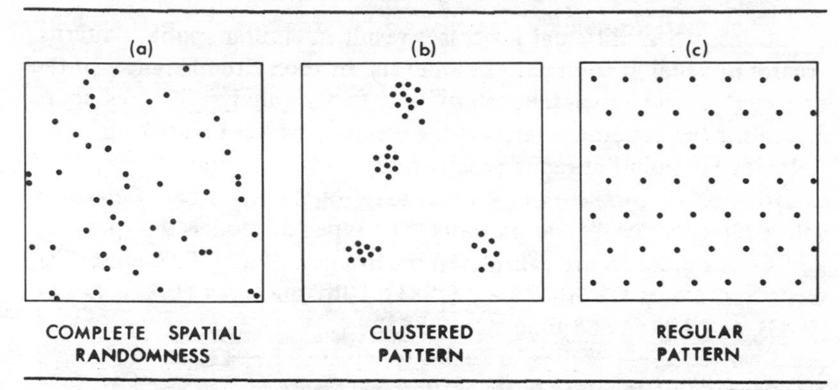

Figure 1.2 Fundamental Types of Patterns: (a) Complete Spatial Randomness (CSR), (b) Clustered Pattern, and (c) Regular Pattern

pattern is CSR produced by a homogeneous planar Poisson point process. This allows for the exploration of a set of data, often leading to the formulation of other geographically relevant hypotheses. Whether or not the initial hypothesis relating to CSR is rejected, a description of the pattern remains. The CSR model is mathematically described by the Poisson probability distribution given in section 2.1.

Classes of patterns can be recognized using CSR as the benchmark. *Clustered patterns* are those in which the points are significantly more grouped in the study area than they are in CSR (see the example in Figure 1.2b), whereas *regular patterns* (sometimes also called uniform or dispersed patterns) are those in which the points are more spread out over the environment than they would be in CSR (see Figure 1.2c). Clustered and regular patterns can arise as the result of changing either or both of the conditions of the homogeneous planar Poisson point process.

A major way of changing the uniformity condition of this model is to turn the homogeneous study area into a heterogeneous one. *Environmental heterogeneity* implies that some locations in the study area are less likely to receive a point than other locations, or might even be prohibited from receiving a point. We would expect to find more points in the favored environmental parts of the study area than elsewhere, thereby producing a clustered pattern.

One way of relaxing the independence assumption is to permit interaction among points: Points may either attract or repulse one another. *Attraction* may result from processes such as agglomeration, association, voluntary or involuntary segregation, and some types of diffusion and competition. In each instance the result is a clustered pattern. Such situations, where different processes result in similar spatial patterns, are not unusual in point pattern analysis. In such circumstances, if the only evidence we have is the pattern itself, further analysis is necessary to determine the conditions responsible for the observed clustering.

Instead of points attracting each other, in some circumstances—such as diffusion or competition—points may repel each other. *Repulsion* will likely produce regular patterns. The types of models that produce non-CSR outcomes are considered by Haggett et al. (1977: chap. 13), Getis and Boots (1978), Ripley (1981), Cliff and Ord (1981), Diggle (1983), and Upton and Fingleton (1985).

1.4 Procedures Involved in Point Pattern Analysis

In section 1.2 we suggested that we may analyze either the dispersion or arrangement characteristics of a point pattern and that the analytical techniques would be presented in terms of which of these characteristics is considered. Regardless of the techniques employed, the general procedures are essentially the same and follow the traditional logical procedures used in hypothesis testing [see Boots and Getis (1977) for a formal presentation of these procedures]. We begin by specifying a null hypothesis, H_0. The null hypothesis in the examples that follow is always the same: It is that the pattern under investigation is CSR resulting from a homogeneous planar Poisson point process. The most simple and general alternative or "research" hypothesis, H_1, is that the pattern is not CSR. Even when one has a preconceived notion of the processes that generated the pattern, it is often a useful starting point to test a null hypothesis relating to CSR. Diggle (1983: 5) suggests three reasons for this. First, if the null hypothesis of CSR is not rejected, further formal statistical analysis is not warranted. Second, as indicated in section 1.3, a null hypothesis of CSR provides a dividing hypothesis between clustered and regular patterns. Finally, even when we anticipate that a null hypothesis of CSR will be rejected, the results of the test can be used as an aid to formulating new null hypotheses.

In such circumstances, the research hypothesis chosen may emphasize specific dependence among observations. Such a hypothesis of dependence might arise from one of the many situations in which like objects are drawn together or evened out. For example, the location of earthquakes in a region can be modeled as a series of clusters, and the location of towns in a rural area can be represented by a dispersion model.

Alternatively, the research hypothesis may concern the heterogeneity of point locations. Such a hypothesis might be used when environmental conditions such as soil types and water and food availability help or hinder location of plant and animal life and human settlement in certain areas.

Once the hypotheses are formulated, statistics are computed using information collected from the map pattern and these are evaluated in terms of the likelihood of their occurrence under the assumptions of the null hypothesis. In all examples considered here we use a significance level of $\alpha = 0.05$. This procedure leads to either the acceptance or rejection of the null hypothesis. Acceptance of the null hypothesis means that the point pattern under investigation is not significantly different from CSR. Rejection of the null hypothesis indicates that the point pattern is significantly different from CSR. Even if the research hypothesis is not specific in terms of the nature of the difference, we are usually able to suggest whether the pattern is a clustered or a regular one. When the null hypothesis is rejected, it is common to formulate a new one that may specify the operation of particular processes leading to the particular type of pattern indicated. Such hypotheses, not discussed in this book, can then be tested in ways similar to those employed here to test the null hypothesis of CSR.

In the next chapter we begin our discussion of particular techniques of point pattern analysis. First, we discuss dispersion as it is commonly studied, by means of the use of quadrat units. In Chapter 3 the discussion of dispersion continues, but here distance methods are introduced. In Chapter 4 we cover a variety of techniques that are used to study patterns when arrangement is the focus. Finally, the summary in Chapter 5 contains advice on choosing one approach rather than another.

The probelms discussed in this monograph can be programmed for solution on computers. We decided not to include programs here because they were written in several languages for different types of computer installations. Readers interested in these programs are invited to contact either of the authors.

2. MEASURES OF DISPERSION: QUADRAT METHODS

As stated in section 1.2, techniques of point pattern analysis can be grouped into two classes; those that examine the location of the points relative to the study area and those that examine the location of points relative to each other. We call the former class *measures of dispersion*, and the latter *measures of arrangement*. The measures of dispersion may be divided further into two classes; quadrat methods and distance methods. The former are the subject of this chapter, and distance methods are discussed in Chapter 3. Measures of arrangement are examined in Chapter 4.

Quadrat is the name given to a sampling area of any consistent shape and size. Quadrats may be located in a study area as isolated individuals or in blocks of contiguous individuals. Techniques involving the former are described in section 2.1; those involving the latter are examined in section 2.2.

2.1 Scattered Quadrats

The procedure described in this section was originally developed by plant ecologists (Greig-Smith 1964) who were concerned with examining point patterns where both N and A were large (often $N > 1000$) and sometimes unknown.

As an example of this approach, consider Figure 2.1, which shows the location of settlements of populations of 300 or greater in 1971 in the Canadian province of Saskatchewan between the border with the United States and latitude 54 degrees north. Here the quadrats are circular. We will discuss below the selection of the appropriate shape, size, and other characteristics of quadrats. In Figure 2.1 35 quadrats have been placed over the study area according to the CSR assumptions. By this we mean that the positions of the centers of the quadrats are selected so that each location in the study area has an equal chance of receiving a quadrat center and that the selection of a location for a quadrat center in no way influences the selection of a location for other quadrat centers. Recall that these are the same conditions used to generate a CSR point pattern (see section 1.3). In practice, the positions of the quadrat centers can be obtained by overlaying an (X, Y) coordinate grid on the point pattern and randomly selecting pairs of (X, Y) coordinates. These pairs of coordinates become the quadrat centers. Once the quadrats have been located we record the number of points, x, in each quadrat. We omit from subsequent analysis the 5 quadrats that are truncated by the boundary, leaving 30 quadrats. Some points may appear to fall on the boundary of a quadrat; however, since points are considered dimensionless such points are recorded as being inside the quadrat. We use this information to construct a table in which we record the frequency of occurrence of a quadrat with a given number of points (see columns 1 and 2 of Table 2.1). Next we determine what the frequencies would look like if the H_0 (of a CSR pattern) was correct. This involves calculating a set of expected frequencies. To do this we must first calculate $p(x)$, the probability of finding a quadrat with x points in a CSR pattern. This probability is given by the Poisson probability distribution, which is

TABLE 2.1 Quadrat Analysis of Towns in Southern Saskatchewan

[1] Number of Points per Quadrat x	[2] Observed Frequency O_i		[3] Probability of a Quadrat with x Points	[4] Expected Frequencies E_i		[5] $(O_i - E_i)^2 / E_i$
0	3	6	0.0743	2.23	8.02	0.509
1	3		0.1931	5.79		
2	8		0.2510	7.53		0.029
3	10		0.2176	6.53		1.844
4	4	6	0.1414	4.24	7.92	0.465
5	1		0.0735	2.21		
6	0		0.0319	0.96		
7	0		0.0118	0.35		
8	1		0.0039	0.12		
>8	0		0.0015	0.04		
Total	30		1.0000	30		2.847

$$p(x) = (e^{-\lambda} \lambda^{x}) / x! \qquad \text{for } x = 0, 1, 2, \ldots \qquad [1]$$

where:

λ is the expected number of points per sample area. This value may be estimated by the mean number of points per quadrat. e is the mathematical constant 2.718282.

For Figure 2.1 the total number of points recorded in the 30 complete quadrats is 78 so that we estimate λ as $78/30 = 2.6$. Thus to obtain the probability $p(x)$ when $x = 0$ (that is, the probability of an empty quadrat) in a CSR pattern we substitute $x = 0$ and $\lambda = 2.6$ into equation 1, noting that $0! = 1$, so that

$$p(x) = [\, e^{-2.6} (2.6)^0] \;/\; 0!$$

$$= e^{-2.6}$$

$$= 0.0743.$$

For $x = 1$

$$p(x) = [e^{-2.6} \; (2.6)^1] \;/\; 1!$$

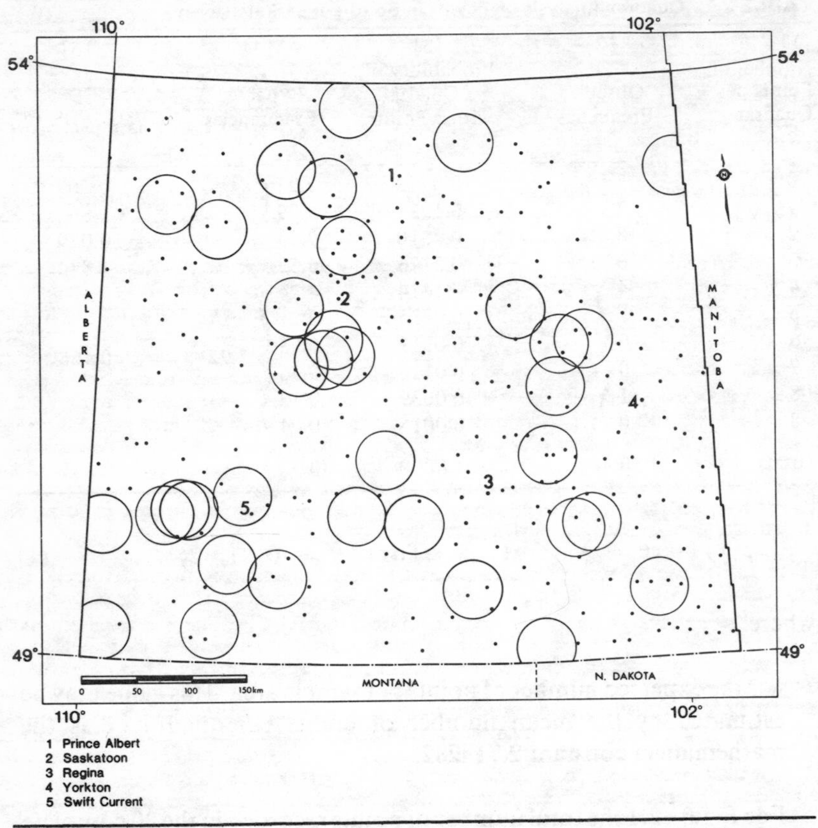

Figure 2.1 Location of Settlements of 300 or Greater Population in 1971 in Southern Saskatchewan

$$= e^{-2.6} \; 2.6$$

$$= 0.1931,$$

and for $x = 2$

$$p(x) = [e^{-2.6} \; (2.6)^2] \; / \; 2!$$

$$= 0.2510.$$

In a similar way the values of p(x) for $x = 3$ through 8 can be obtained.[1] Although there are no quadrats with more than eight points, the probability of $p(x)$ for x greater than eight is not zero and so we calculate $p(x > 8)$. Since all the probabilities must sum to one, this value is obtained from the equation

$$p(x > 8) = 1 - \sum_{x=0}^{8} p(x) \qquad [2]$$

In this case $p(x > 8) = 1 - 0.9985 = 0.0015$. When we have obtained all the probabilities of the occurrence of a quadrat with x points (see column 3 of Table 2.1), we obtain the expected frequencies of quadrats with x points by multiplying the appropriate probability by the total number of quadrats. In this example, the number of quadrats is 30 so that the expected frequency of empty quadrats is (30) (0.0743) = 2.23, the expected number with one point is (30) (0.1931) = 5.79, with two points is (30) (0.2510) = 7.53, and so on. The complete set of these expected frequencies is shown in column 4 of Table 2.1. Our test of the H_0 consists of comparing the expected frequencies with the observed frequencies in column 2. The procedure most often used is the chi-square one sample, goodness-of-fit test. This test uses the following formula to obtain a chi-squared statistic, X^2, to compare the two sets of frequencies:

$$X^2 = \sum_{i=1}^{K} (O_i - E_i)^2 / E_i \qquad [3]$$

where:

O_i is the observed frequency in the ith category
E_i is the expected frequency in the ith category
K is the number of categories.

The value of K used in the test depends on the expected frequencies. Although there is some disagreement on this matter, a conservative suggestion is that the expected frequencies in any category should be at least five. If some expected frequencies are smaller than five, adjacent values are combined until five is reached or exceeded. Thus in Table 2.1 the expected frequencies for $x \leq 1$ and for $x \geq 4$ are combined into single categories. The corresponding observed frequencies are grouped accordingly. This leaves us with four categories ($K = 4$). Column 5 of Table 2.1 lists the individual values of equation 3, which are summed to obtain

$X^2 = 2.85$. This value is compared to the value of chi-square (χ^2) obtained from statistical tables. To use the statistical tables, the degrees of freedom (*df*) for the test must be computed. The number of *df* is equal to the number of comparisons made minus one, $(K - 1)$, and an additional *df* is lost for each value estimated in order to obtain the expected frequencies. In this example, because we need to estimate λ, the appropriate df for the test are $(4\text{-}1\text{-}1) = 2$. Values of X^2 that fall below the value of χ^2 for the predetermined significance level (in our case, $\alpha = 0.05$) indicate that we cannot reject the H_0. Values of X^2 greater than χ^2 mean that the H_0 may be rejected. From statistical tables with $df = 2$, $\chi^2 = 5.99$. Since our computed X^2 is 2.85, the H_0 is accepted. Therefore, we conclude that the observed pattern of towns is not significantly different from CSR.

2.2 The Variance/Mean Ratio

Consider for a moment what would have happened if our value of X^2 in the previous example was such that we would reject the H_0. Assuming that our H_1 is that the pattern is not CSR, can we say that the observed pattern is a regular or a clustered one? Strictly speaking, the answer is no. However, we can use the data collected already to create another test that will enable us to reach an answer. This is the variance of the number of points per quadrat, V. This may be calculated using

$$V = \frac{\sum_{x} (x - \lambda)^2 fx}{n} \qquad [4]$$

where

n is the number of quadrats

fx is the observed frequency of x.

In a Poisson probability distribution, the value of λ is expected to equal V. Thus comparisons of the V and λ derived from our data provide a convenient test of the hypothesis of CSR. If we have a regular pattern, each quadrat will contain a similar number of points and thus the value of V will be less than λ. On the other hand, for a clustered pattern there will be many quadrats with few or no points, corresponding to the spaces between the individual clusters in the pattern and a few quadrats, located in the clusters, which contain a relatively large number of points. Such a situation will generate a value of V in excess of λ. This idea can be expressed more formally and used to provide an alternative to the chi-square test discussed earlier. This alternative test involves calculating a t statistic using the following equation

$$t = \frac{(V - \lambda)}{[2/(n-1)]^{1/2}} . \quad [5]$$

The calculated value of t may be compared with the value of t from statistical tables for the appropriate significance level. The degrees of freedom, df, for the test are $(n-1)$. If the absolute calculated value of t does not exceed the value from the tables, we cannot reject the H_0 of CSR. However, if the absolute calculated value t exceeds the value from the tables, we may reject the H_0. In such cases, if the calculated value of t is positive, V must be greater than λ and a clustered pattern is indicated, whereas a negative value of t arises because V is less than λ, suggesting a regular pattern. For the pattern in Figure 2.1, V is 2.51 so that

$$t = (2.51 - 2.60) / (2/29)^{1/2}$$

$$= -0.35$$

and $df = 29$. The value of t from statistical tables is 2.05, so that we retain the H_0 confirming the result we obtained above for the chi-square test.

2.3 Contiguous Quadrats

The "scattered quadrat" method described earlier has been used only rarely in geography. This is mainly because the patterns examined by geographers usually are much more restrictive in terms of both N and A. In such circumstances, in order to obtain sufficient points for the analysis, the size of the quadrats relative to A must be increased. This increases the likelihood of quadrats overlapping (as happened in the example in Figure 2.1). When quadrats overlap they are no longer independent of each other, thereby violating the independence condition involved in generating CSR (see section 1.3). In order to avoid this, applications of quadrat analysis in geography usually employ a grid of quadrats that are superimposed on the study area. This procedure does not violate the independence condition and also has the advantage that most of the points in the pattern are used in the analysis. However, there are still a number of problems that have to be resolved. First, what should be the shape of the quadrats? Although various packable shapes such as triangles, hexagons, and rectangles could be used, square quadrats are used almost invariably. This is because they are easy to construct and, if desired, may be aggregated easily into larger units.

The choice of quadrat size is a more difficult issue. First, there is a tendency for large quadrats to produce a situation where there are approximately the same large number of points in all of the quadrats thus biasing the result toward an even pattern. However, if the quadrats are too small they may effectively subdivide any clusters present in the pattern leading to a situation with only two possible outcomes for a quadrat, a single point or no point, producing a bias toward CSR. Obviously, some compromise must be found between these extremes, and a number of studies have attempted to do so (these are discussed in Pielou 1969: 100-104; Upton and Fingleton 1985: 31). One useful rule of thumb suggested by some of these studies is that the appropriate size of a quadrat can be approximated as twice the size of the area per point. Thus if we use square quadrats, the length of a side, Q, would be in the order of $\sqrt{(2A/N)}$. However, it should be stressed that this is only a rule of thumb and other factors may influence the final decision on quadrat size. In fact, it is not unusual to select a number of different quadrat sizes and repeat the tests. In such circumstances square quadrats are advantageous because they can be combined easily. Such aggregation means, however, that the tests at the different scales may well correlate with each other.

Another variable associated with a grid of quadrats is its orientation. This is not important if we are testing an H_0 of CSR since the two conditions associated with the process that generates CSR ensure that there will be no directional trends in the pattern.

2.4 Spatial Autocorrelation

Perhaps the most severe limitation of quadrat analysis is that by summarizing the point pattern as a set of frequencies it loses the spatial dimension of the pattern (Dacey 1966a). This means that quite different patterns, such as the three in Figure 2.2, when summarized by quadrats are reduced to the same set of frequencies with the effect that any quadrat analysis performed on them produces identical results. This illustrates that quadrat analysis is insensitive to the spatial arrangement of the quadrats containing varying numbers of points and that if such characteristics are to be accounted for, an additional statistical procedure is required. Most often this involves an examination of the extent of spatial autocorrelation (Odland 1987) in the values in the quadrats.

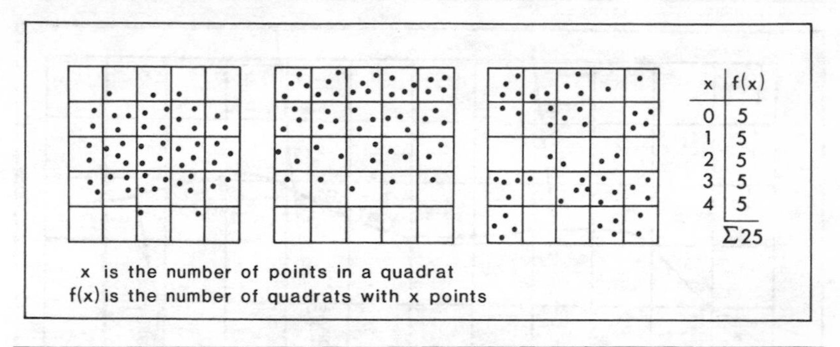

Figure 2.2 Different Point Patterns that Produce the Same Set of Quadrat Frequencies

Simply defined, spatial autocorrelation is a measure of the correlation among neighboring observations in a pattern. If the observations are numbers of points in individual quadrats, then spatial autocorrelation will determine the extent to which values in neighboring quadrats are correlated. No spatial autocorrelation implies that there is no correlation between neighboring values. This is the situation we would expect for a CSR pattern since one of the conditions involved in its generation is that of independence among points (see section 1.3). Thus, in the case of an H_0 of CSR, even if the quadrat analysis leads to the acceptance of the H_0, we should in addition perform some test to find out if we have spatial autocorrelation in the quadrat values.

To illustrate some of these concerns, consider Figure 2.3, which shows the location of Separate (Roman Catholic) schools in metropolitan Toronto, excluding the eastern municipality of Scarborough in 1976. Here the usual grid of square quadrats is used. Each quadrat has a side of 2.32 kms. This somewhat strange value arises because this size corresponds to a quadrat of side 2.5 inches on the map used in the analysis. Nevertheless, since $A = 453.26$ sq. kms and $N = 134$, the size chosen is of the same order as that suggested by the rule of thumb given above since $Q = \sqrt{(2A/N)} = 2.60$ kms. Since the study area has an irregular boundary the numbers of rows and columns in the grid are not equal and the grid is positioned so that as few quadrats as possible truncate the boundary (and are consequently lost to the analysis). Once the grid is in position we proceed as before by recording the number of points in each quadrat and constructing a table of these frequencies (see columns 1 and 2 of Table 2.2). These frequencies are also shown in Figure 2.4a. Note we do not include those quadrats that overlap the study area boundary. In this way 18 quadrats containing a total of 15

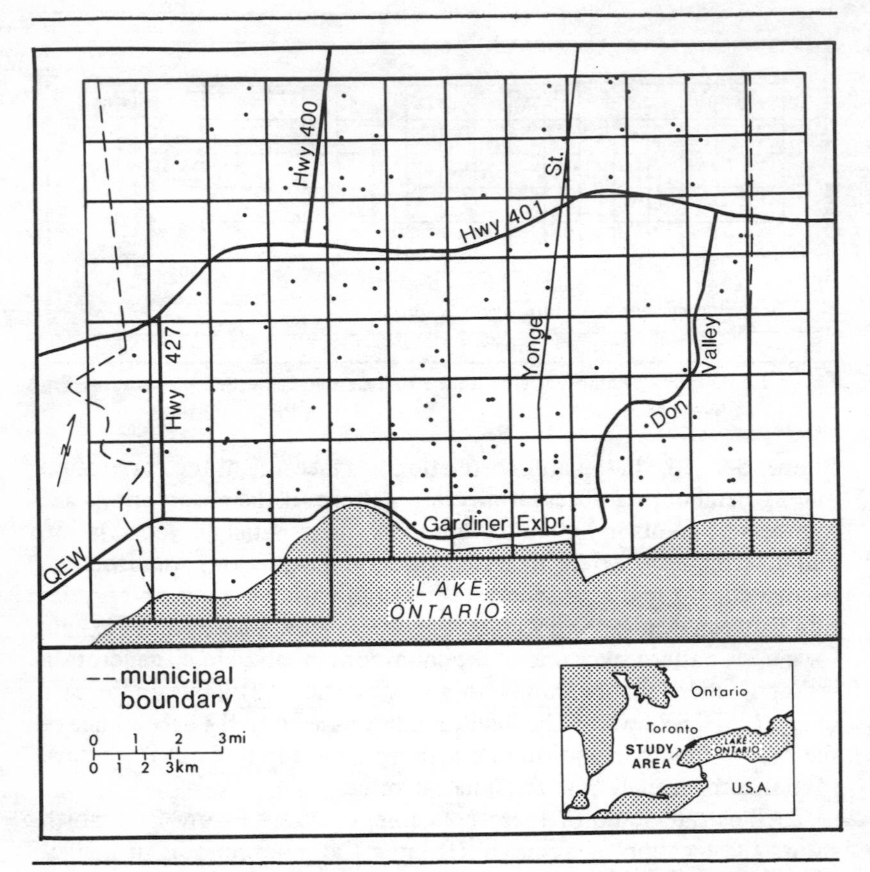

Figure 2.3 Location of Separate Schools in Metropolitan Toronto (Excluding the Municipality of Scarborough) in 1976

points are lost from the analysis, leaving the 72 quadrats shown in Figure 2.4a. As before, the probability of a quadrat containing x points in a CSR pattern is obtained from equation 1 (see column 3 of Table 2.2). In this case, since both N and A are known, we estimate λ from $(N/A)Q^2 = (134/453.26)(2.32)^2 = 1.595$. These probabilities are multiplied by the number of quadrats (72) to obtain the expected frequency of quadrats with x points (see column 4 of Table 2.2). Equation 3 is used to obtain X^2 which is 2.16 with df = (5-1-1) = 3. From the chi-square tables the value of χ^2 with $df = 3$ and $\alpha = 0.05$ is 7.82. Thus the H_0 cannot be rejected; at the same time it cannot be concluded that we have CSR until we test for the extent of spatial autocorrelation in the quadrat values.

TABLE 2.2 Quadrat Analysis of Separate Schools in Metropolitan Toronto (Excluding the Borough of Scarborough)

[1] Number of Points per Quadrat x	[2] Observed Frequency O_i	[3] Probability of a Quadrat with x Points	[4] Expected Frequencies E_i	[5] $(O_i - E_i)^2 / E_i$
0	17	0.2029	14.61	0.391
1	23	0.3236	23.30	0.004
2	15	0.2581	18.58	0.690
3	9	0.1372	9.88	0.078
4	4	0.0547	3.94	
5	3	0.0175	1.26	
6	0 } 8	0.0046	0.33 } 5.63	0.998
7	0	0.0011	0.08	
8	1	0.0002	0.01	
>8	0	0.0001	0.01	
Total	72	1.0000	72.00	2.161

A variety of procedures has been developed to measure spatial autocorrelation (see Cliff and Ord 1981: chapts. 1 and 2; Odland 1987) but one that is appropriate here is the I statistic developed by Moran (1950):

$$I = \frac{\frac{n}{2a} \sum_{i=1}^{n} \sum_{\substack{j=1 \\ i \neq j}}^{n} \delta_{ij}(x_i - \bar{x})(x_j - \bar{x})}{\sum_{i=1}^{n} (x_i - \bar{x})^2} \qquad [6]$$

where:

n is the number of quadrats

δ_{ij} is a measure of the contiguity between quadrat i and quadrat j. $\delta_{ij} = 1$ if i and j are contiguous, $\delta_{ij} = 0$, otherwise. Contiguity may be defined as having at least one edge in common (rook's case), at least one vertex in common (bishop's case), or at least one edge or one vertex in common (queen's case)

x_i is the number of points in quadrat i

0	3	0	3	0	0	2	4	3	1
1	0	2	3	1	1	2	1	0	2
0	1	3	1	2	0	2	0	0	2
1	1	1	4	1	1	1	0	1	2
0	2	0	3	3	0	0	1	1	0
1	2	1	2	5	3	2	0	1	1
1	4	2	0	4	8	5	2	5	3
1	2								

(a)

1	2	3	4	5	6	7	8	9	10
11	12	13	14	15	16	17	18	19	20
21	22	23	24	25	26	27	28	29	30
31	32	33	34	35	36	37	38	39	40
41	42	43	44	45	46	47	48	49	50
51	52	53	54	55	56	57	58	59	60
61	62	63	64	65	66	67	68	69	70
71	72								

(b)

Figure 2.4 Quadrat Analysis of Separate Schools in Metropolitan Toronto: (a) Frequencies of Occurrence of Schools, and (b) Identification Numbers of Quadrats

x_j is the number of points in quadrat j
$\bar{x}$ is the mean number of points per quadrat

$$2a \quad \text{is equal to} \sum_{\substack{i=1 \\ i \neq j}}^{n} \sum_{j=1}^{n} \delta_{ij}$$

The calculated value of I can be compared with the set of all possible values I can take on if the values of x_i (number of points per quadrat) are repeatedly randomly permuted over the set of quadrats. There are $n!$ such values and the expected value, E(I), is given by

$$\mathrm{E}(I) = -(n-1)^{-1} \qquad [7]$$

The difference between the observed and expected values of I can be evaluated by a normally distributed statistic, z, of the form

$$z = [I - \mathrm{E}(I)] \ / \ \sqrt{\mathrm{var}(I)} \qquad [8]$$

where

$$\mathrm{var}(I) = \mathrm{E}\ (I^2) - [\mathrm{E}\ (I)]^2 \qquad [9]$$

and

$$\mathrm{E}\ (I^2) = \{n\ [4\ a\ (n^2 - 3\ n + 3) - 8\ (a + d)n + 12a^2]$$
$$-b_2\ [4\ a(n^2 - n) - 16\ (a + d)n + 24a^2]\} \ / \ [4a^2(n - I)$$
$$(n\text{-}2)\ (n\text{-}3)] \qquad [10]$$

where:

$$d = 1/2 \sum_{i=1}^{n} L_i\ (L_i - 1) \qquad [11]$$

where:

L_i = the number of quadrats contiguous to quadrat i

and

$$b_2 = m_4 / m_2^2 \qquad [12]$$

where:

$$m_2 = \sum_{i=1}^{n} (x_i - \bar{x})^2 / n$$

and

$$m_4 = \sum_{i=1}^{n} (x_i - \bar{x})^4 / n \quad .$$

In this test the H_0 is no spatial autocorrelation, which implies that the values are randomly distributed over the quadrats. However, if I is found to be significantly greater than E(I) the pattern of quadrat values is said to display positive spatial autocorrelation—that is, similar values are located in proximity to each other. If I is significantly less than E(I), we have negative spatial autocorrelation implying that like values are close to unlike ones.

To illustrate this procedure, consider again the grid of quadrats and their associated values shown in Figure 2.4a. Begin by calculating, $\bar{x}$, the average number of points per quadrat, which appears in both the numerator and denominator of equation 6. Since there are 119 schools and 72 quadrats, $\bar{x}$ = 1.653.

Next consider the denominator of equation 6. This instructs us to calculate the sum over all quadrats of the squared deviation between the number of points, x_i, in quadrat i and the average number, $\bar{x}$. The steps involved are given in Table 2.3. If the individual quadrats are numbered as in Figure 2.4b, columns 1 and 2 of Table 2.3 give the quadrat indentification number, i, and the number of points in the quadrat, x_i, respectively. Column 3 of Table 2.3 gives the difference between x_i and

TABLE 2.3 Calculation of $\sum_{i=1}^{n} (x_i-\bar{x})^2$ in Equation 6

[1]	[2]	[3]	[4]	[5]
Quadrat i	Number of Points in Quadrat x_i	$(x_i-\bar{x})$	$(x_i-\bar{x})^2$	$(x_i-\bar{x})^4$
1	0	−1.653	2.732	7.464
2	3	1.347	1.814	3.291
3	0	−1.653	2.732	7.464
4	3	1.347	1.814	3.291
5	0	−1.653	2.732	7.464
6	0	−1.653	2.732	7.464
7	2	9.347	0.120	0.014
8	4	2.347	5.508	30.338
9	3	1.347	1.814	3.291
10	1	−0.653	0.426	0.181
11	1	−0.653	0.426	0.181
12	0	−1.653	2.732	7.464
13	2	0.347	0.120	0.014
14	3	1.347	1.814	3.291
15	1	−0.653	0.426	0.181
16	1	−0.653	0.426	0.181
17	2	0.347	0.120	0.014
18	1	−0.653	0.426	0.181
19	0	−1.653	2.732	7.464
20	2	0.347	0.120	0.146
21	0	−1.653	2.732	7.464
22	1	−0.653	0.426	0.181
23	3	1.347	1.814	3.291
24	1	−0.653	0.426	0.181
25	2	0.347	0.120	0.014
26	0	−1.653	2.732	7.464
27	2	0.347	0.120	0.014
28	0	−1.653	2.732	7.464
29	0	−1.653	2.732	7.464
30	2	0.347	0.120	0.014
31	1	−0.653	0.426	0.181
32	1	−0.653	0.426	0.181
33	1	−0.653	0.426	0.181
34	4	2.347	5.508	30.338
35	1	−0.653	0.426	0.181
36	1	−0.653	0.426	0.181
37	1	−0.653	0.426	0.181
38	0	−1.653	2.732	7.464
39	1	−0.653	0.426	0.181
40	2	0.347	0.120	0.014
41	0	−1.653	2.732	7.464

Continued

TABLE 2.3 Continued

[1] Quadrat i	[2] Number of Points in Quadrat x_i	[3] $(x_i - \bar{x})$	[4] $(x_i - \bar{x})^2$	[5] $(x_i - \bar{x})^4$
42	2	0.347	0.120	0.014
43	0	−1.653	2.732	7.464
44	3	1.347	1.814	3.291
45	3	1.347	1.814	3.291
46	0	−1.653	2.732	7.464
47	1	−0.653	0.426	0.181
48	1	−0.653	0.426	0.181
49	1	−0.653	0.426	0.181
50	0	−1.653	2.732	7.464
51	1	−0.653	0.426	0.181
52	2	0.347	0.120	0.014
53	1	−0.653	0.426	0.181
54	2	0.347	0.120	0.014
55	5	3.374	11.202	125.485
56	3	1.347	1.814	3.291
57	2	0.347	0.120	0.014
58	0	−1.653	2.732	7.464
59	1	−0.653	0.426	0.181
60	1	−0.653	0.426	0.181
61	1	−0.653	0.426	0.181
62	4	2.347	5.508	30.338
63	2	0.347	0.120	0.014
64	0	−1.653	2.732	7.464
65	4	2.347	5.508	30.338
66	8	6.346	40.284	1622.801
67	5	3.347	11.202	125.485
68	2	0.347	0.120	0.014
69	5	3.347	11.202	125.485
70	3	1.347	1.814	3.291
71	1	−0.653	0.426	0.181
72	2	0.347	0.120	0.014

the mean number of points per quadrat, $\bar{x}$, whereas column 4 gives the square of this value. The sum of the individual values in column 4 of Table 2.3 is the term

$$\sum_{i=1}^{n} (x_i - \bar{x})^2$$

of equation 6. This value is equal to 170.29.

Now consider the term

$$\sum_{i=1}^{n} \sum_{\substack{j=1 \\ i \neq j}}^{n} \delta_{ij} (x_i - \bar{x}) (x_j - \bar{x})$$

of equation 6. This involves evaluating $\delta_{ij} (x_i - x) (x_j - x)$ for all pairs of quadrats, i,j ($i \neq j$). Note, however, that the measure of contiguity, δ_{ij}, between quadrats i and j can take on only two values, 1 and 0. The value 0 occurs if two quadrats are not contiguous. When this occurs the corresponding value of $\delta_{ij}(x_i - \bar{x}) (x_j - \bar{x})$ for that pair of quadrats is zero and contributes nothing to the overall sum. Thus we need only consider those pairs of quadrats that are contiguous and for which δ_{ij} equals 1. Inspection of Figure 2.4b reveals that there are 252 such contiguous pairs. Those pairs for the top row only of the grid of quadrats in Figure 2.4b are given in columns 1 and 2 of Table 2.4. Columns 3 and 4 of Table 2.4 show the corresponding numbers of points, x_i, x_j, in the contiguous quadrats, whereas columns 5 and 6 show the differences between these values and $\bar{x}$, the mean number of points per quadrat. Finally, column 7 of Table 2.4 gives the product term $(x_i - \bar{x})(x_j - \bar{x})$. If this value is also calculated for the remaining 224 pairs of contiguous quadrats in Figure 2.4b and added to those values in column 7 of Table 2.4 the result is the term

$$\sum_{i=1}^{n} \sum_{\substack{j=1 \\ i \neq j}}^{n} \delta_{ij} (x_i - \bar{x}) (x_j - \bar{x}).$$

For this example this value is equal to 157.104.

Finally, in order to complete the calculation of I in equation 6 we need to calculate $n/2a$, where n, the number of quadrats, is 72 and $2a$ is the number of pairs of contiguous quadrats, which is 252. Thus $n/2a$ = 0.286. Inserting the above values in equation 6 yields

$$I = (0.286)\ (157.104)\ /\ 170.290$$

$$= 0.2639.$$

From equation 7, $E(I) = -(72 - 1)^{-1} = -0.0141$. The values of L_i in equation 11 can be obtained by examining Figure 2.4b. Thus L_1, the

TABLE 2.4 Calculation of $\sum_{i=1}^{n} \sum_{\substack{j=1 \\ i \neq j}}^{n} \delta_{ij} (x_i - \bar{x})(x_j - \bar{x})$ **in Equation 6**

[1] Quadrat i	[2] j	[3] Number of Points in Quadrat x_i	[4] x_j	[5] $(x_i-\bar{x})$	[6] $(x_j-\bar{x})$	[7] $(x_i-\bar{x})(x_j-\bar{x})$
1	2	0	3	–1.653	1.347	–2.227
1	11	0	1	–1.653	–0.653	1.079
2	1	3	0	1.347	–1.653	–2.227
2	3	3	0	1.347	–1.653	–2.227
2	12	3	0	1.347	–1.653	–2.227
3	2	0	3	–1.653	1.347	–2.227
3	4	0	3	–1.653	1.347	–2.227
3	13	0	2	–1.653	0.347	–0.574
4	3	3	0	1.347	–1.653	–2.227
4	5	3	0	1.347	–1.653	–2.227
4	14	3	3	1.347	1.347	1.814
5	4	0	3	–1.653	1.347	–2.227
5	6	0	0	–1.653	–1.653	2.732
5	15	0	1	–1.653	–0.653	1.079
6	5	0	0	–1.653	–1.653	2.732
6	7	0	2	–1.653	0.347	–0.574
6	16	0	1	–1.653	–0.653	1.079
7	6	2	0	0.347	–1.653	–0.574
7	8	2	4	0.347	2.347	0.814
7	17	2	2	0.347	0.347	0.120
8	7	4	2	2.347	0.347	0.814
8	9	4	3	2.347	1.347	3.161
8	18	4	1	2.347	–0.653	–1.533
9	8	3	4	1.347	2.347	3.161
9	10	3	1	1.347	–0.653	–0.880
9	19	3	0	1.347	–1.653	–2.227
10	9	1	3	–0.653	1.347	–0.880
10	20	1	2	–0.653	0.347	–0.227

number of quadrats contiguous to quadrat 1 is 2, $L_2 = L_3 = L_4 = L_5 = L_6 = L_7 = L_8 = L_9 = 3$, $L_{10} = 2$, $L_{11} = 3$, $L_{12} = 4$, and so on. Inserted in equation 11 these values yield $d = 329$.

For equation 12, m_2 is equal to the term

$$\sum_{i=1}^{n} (x_i - \bar{x})^2$$

already calculated in the computation of I, divided by n. Thus m_2 = (170.29/72) = 2.365. Since $(x_i - \bar{x})^4$ equals $[(x_i - \bar{x})^2]^2$, m_4 in equation 12 can be calculated by taking the values in column 4 of Table 2.3, squaring them and summing. The resulting values are given in column 5 of Table 2.3 and yield m_4 = 2281.488/72 = 31.687. Thus in equation 12, b_2 = $(31.687)/(2.365)^2$ = 5.665. Inserting this value of b_2 together with the values of n = 72, a = 126, and d = 329 into equation 10 gives $E(I^2) = 0.0074$. This, in turn, inserted in equation 9 gives var(I) = 0.0072. Finally, using equation 8, we get $z = [0.2639 + 0.0141]/\sqrt{(0.0072)} = 3.276$. Since the value for z from the tables of the Normal distribution for $\alpha = 0.05$ is 1.96, we reject the H_0 in favor of one that indicates positive spatial autocorrelation. We cannot accept that the pattern of schools is CSR as suggested by quadrat analysis alone. In fact, since the positive spatial autocorrelation results from a clustering of like values, we might hypothesize that the pattern reflects a greater probability of schools occurring in some areas of metropolitan Toronto due to spatial variation in the distribution of Catholic children of school age.

Quadrat analysis has been used in geography to examine a wide variety of phenomena. In human geography these include shop location (Getis 1964; Rogers 1965, 1974; Artle 1965; Sibley 1972; Lee 1974), central place systems (Olsson 1966), settlement patterns (Dacey 1964, 1966a, 1966b, 1968, 1969a-e; Birch 1967), urban growth simulation (Malm et al. 1966) and the diffusion of agricultural innovations (Harvey 1966). Examples of the use of quadrats in physical geography include the study of karst depressions (McConnell and Horn 1972), drumlins (King 1974; Gravenor 1974; Trenhaile 1971) area volcanism (Tinkler 1971), and river channel networks (Oeppen and Ongley 1975). Reviews of this and other works are provided by Thomas (1979, 1981).

3. MEASURES OF DISPERSION: DISTANCE METHODS

In view of the problems and limitations of quadrat techniques, other methods of measuring dispersion properties of point patterns have been developed. The largest set of such alternative techniques is that which is collectively known as distance methods. These techniques analyze a point pattern by way of statistics that are calculated using characteristics of the distances separating individual points in the pattern.

Like quadrat analysis, distance methods were originally developed by plant ecologists (Greig-Smith 1964; Southwood 1966: 39-43; Pielou 1969: chap. 10). The earliest methods involved examining the average of the distances between each point in the pattern and the closest point to it (Clark and Evans 1954). Such methods are referred to as nearest neighbor analysis and are described for two- and one-dimensional study areas in sections 3.1 and 3.4, respectively. As originally developed, the nearest neighbor techniques were subject to several limitations. Two major limitations, inaccuracy in interpretation in some situations and edge effects, are examined in sections 3.2 and 3.3, respectively. More recently, techniques using the entire distribution of distances between each point and its nearest neighbor have been developed. These procedures are referred to as refined nearest neighbor analysis (Diggle 1979a: 79) and form the topic of section 3.5. Other recent developments involve using other interpoint distances in addition to those between individual points and their nearest neighbors. These are known as second-order procedures and are discussed in section 3.6.

3.1 Nearest Neighbor Analysis in Two Dimensions

The nearest neighbor distance for any point is defined as the distance between it and the nearest other point in the pattern. One of the simplest of such nearest neighbor tests involves selecting a number of points from the pattern at random. As before, this means that each of the points in the pattern has an equal chance of being selected and that the selection of any point in no way influences the selection of other points. In practice, in order to perform this sampling procedure we have to number uniquely each point in the pattern and randomly select individual sample points. We identify the nearest neighbor distance (d_i) for each of the sample points (i), sum these

$$\sum_{i=1}^{n} d_i,$$

and obtain the mean nearest neighbor distance

$$\bar{d} = \sum_{i=1}^{n} d_i / n$$

where n is the number of sampled points. Assuming our usual H_0, d is compared to that which would be expected for a random sample of points from a CSR pattern.

As an example of this approach, consider Figure 3.1, which shows the locations of fires that caused in excess of $1000 damage in 1981 in the city of Kitchener in southern Ontario, Canada. There are 132 points in the pattern. We have selected 50 of these points at random and these are shown by the filled circles in Figure 3.1. Then we find the nearest neighbor distance for each of the sampled points. Five examples of such distances are shown in Figure 3.1. Using these distances, we get a value of $\bar{d} = 0.22$ miles. Clark and Evans (1954) show that the expected value of the average nearest neighbor distance, $E(d_i)$, for a random sample of points from a CSR pattern is approximated by the equation

$$E(d_i) = 0.5 \sqrt{(A/N)}. \quad [13]$$

In this example $A = 52.17$ sq. miles and $N = 132$, so that equation 13 gives $E(d_i) = 0.31$ miles. The observed and expected values may be compared using a normally distributed z statistic of the form

$$z = [\bar{d} - E(d_i)] \; / \; \sqrt{\mathrm{var}\,(\bar{d})} \quad [14]$$

where

$$\mathrm{var}(\bar{d}) = 0.0683 A/N^2. \quad [15]$$

Equation 15 gives $\mathrm{var}(\bar{d}) = 0.000205$; substituting this value into equation 14 yields

$$z = (0.22-0.31) \; / \; \sqrt{0.000205} = -6.58.$$

The value of z from tables of the normal distribution for $\alpha = 0.05$ is 1.96. Since the absolute value of the calculated z is greater than 1.96, we reject the H_0 and accept the H_1. In this example the calculated value of z was negative because $\bar{d} < E(d_i)$. This implies that, on average, the individual points are closer than they would be in a CSR pattern, indicating that we have a clustered pattern. In this example such clustering probably represents a greater likelihood of fires starting in areas of the city with certain environmental characteristics, such as old, crowded, delapidated buildings. On those occasions when the H_0 of CSR is rejected and $\bar{d} > E(d_i)$, a regular pattern is indicated.

For many of the patterns encountered by geographers, N is typically

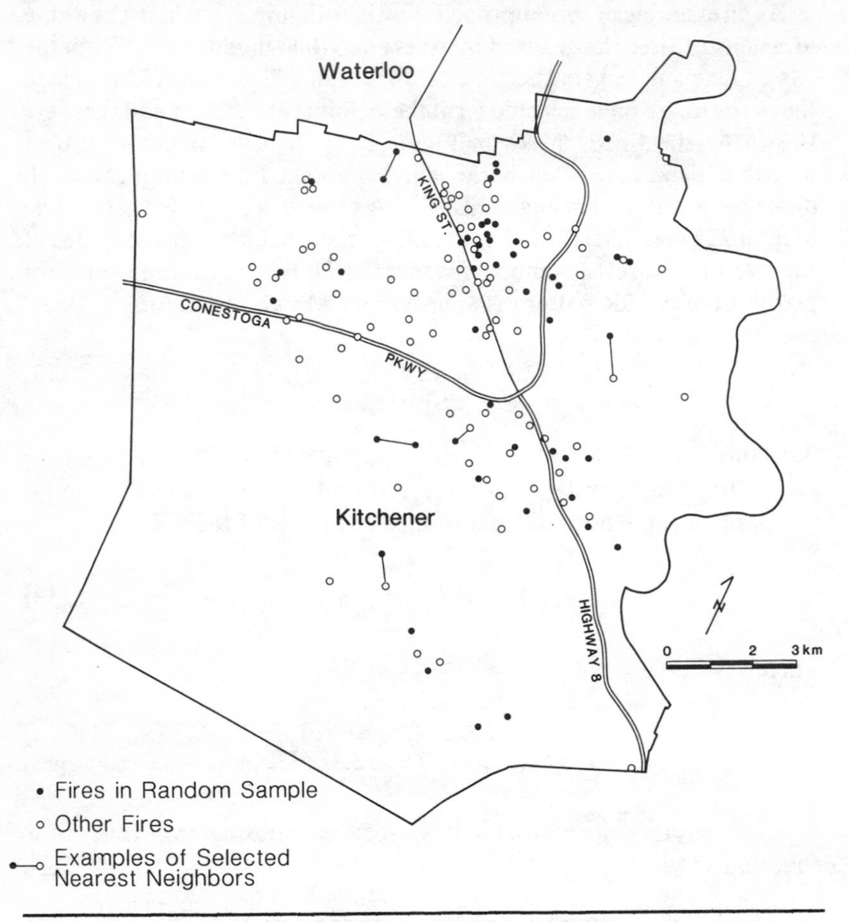

Figure 3.1 Locations of Fires Causing in Excess of $1000 Damage in 1981 in the City of Kitchener, Ontario

quite small and usually $N < 100$. In such circumstances, measuring the d_i's for a random sample of points from the pattern is not very practical. Instead of doing this we may measure the d_i's for all the points in the pattern and use these to calculate $\bar{d}$. When we perform nearest neighbor analysis using all the d_i's, the individual d_i's can no longer be considered mutually independent as they would be for a random sample of points when the size of the sample is small relative to N. Theoretically, this lack of independence is important because it is assumed in Clark and Evans's derivation of equations 13 and 15 (Hsu and Mason 1974; Cliff and Ord 1975). However, Diggle (1976) has shown that this lack of independence has little effect on equations 13 and 15, so that they may also be used when $\bar{d}$ is calculated using all the d_i's.

Nevertheless, this technique of nearest neighbor analysis does have two major inherent problems. These are discussed in the next two sections.

3.2 Higher-Order Neighbor Distances

A problem arises with nearest neighbor analysis when the pattern under study is the result of the operation of more than one distinct process. Figure 3.2 shows a pattern in which individual points form pairs but in which the pairs form a more or less regular pattern (couples on a dance floor, perhaps). Application of the nearest neighbor technique described above would yield a value of $\bar{d} < E(d_i)$, suggesting that we have a clustered pattern and thus revealing only part of the story.

However, if in addition to examining distances to the closest point, we examine the distance between each point and its second nearest neighbor, the regular nature of the spatial locations of the clusters would be identified. Distances other than those between a point and its closest neighbor are often referred to as second-, third-, or *higher-order neighbor distances.* The number of distances we choose to examine will depend on the pattern being studied. However, the form of the test remains the same as that described in equation 14, except that it is necessary to make changes in the values of the constants in equations 13 and 15 depending on the order of the neighbor distances. The values for the constants are given in Table 3.1. Examples of the use of higher-order nearest neighbor analysis include Dacey and Tung (1962), Dacey (1963, 1965), Jones (1971), Trenhaile (1971, 1975), and Tinkler (1971).

3.3 Edge Effects

Another assumption underlying Clark and Evans's derivation of equations 13 and 15 is that the values in these equations relate to the nearest neighbor properties of an *infinite* or unbounded CSR pattern. Depending on the circumstances, application of equations 13 and 15 in the context of the bounded situations encountered in real-world patterns can lead to spurious results. Thus to use the test statistic in equation 14 it is necessary to have some means of compensating for empirical patterns being bounded.

There are four methods that are used to compensate for such edge effects; which method is used depends on both N and the shape of the study area. First, if the study area is a rectangle or a square, several

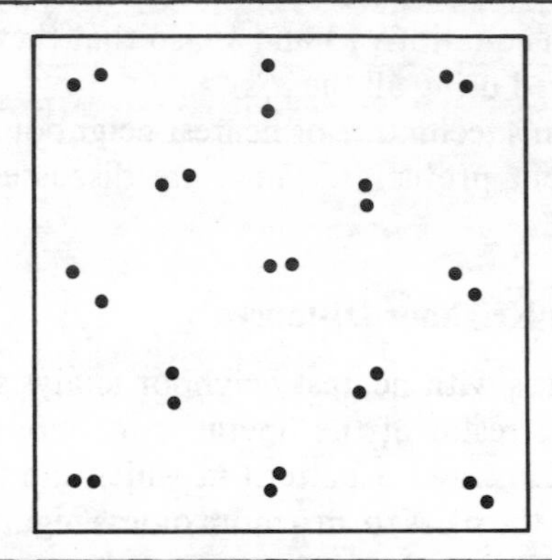

Figure 3.2 Pattern Resulting from the Operation of More than One Process

TABLE 3.1 Values of Constants in Expressions for E (d_i) and Var $(\bar{d})$ for First-Order and Higher-Order Nearest Neighbors in a Random Pattern

Order of Neighbors	$E(d_i) = \gamma_1 \sqrt{A/N}$ γ_1	$\text{var}(\bar{d}) = \gamma_2 A/N^2$ γ_2
1	0.5000	0.0683
2	0.7500	0.0741
3	0.9375	0.0760
4	1.0937	0.0770
5	1.2305	0.0775
6	1.3535	0.0778

SOURCE: Thompson (1956).

studies (Dacey 1975; Ingram 1978; Ripley 1979a,b; Griffith and Amrhein 1983) suggest it is best to convert it to a torus. This involves the joining together the opposite edges of the study area to form a "dough-nut" type of figure over whose surface we measure the required distances. The folding procedure is illustrated in Figure 3.3. In practice, toroidal mapping is achieved by surrounding the study area with identical point patterns as shown in Figure 3.4a. By adopting such a strategy we assume that the same processes responsible for the location of the points in the study area are operating beyond its boundaries. Of course, if the study area does not possess a regular boundary, toroidal mapping cannot be used.

Second, an alternative to toroidal mapping to overcome the boundary problem is to include in the calculation of $\bar{d}$ only those values of d_i that are less than the distance between i and the boundary of the study area.

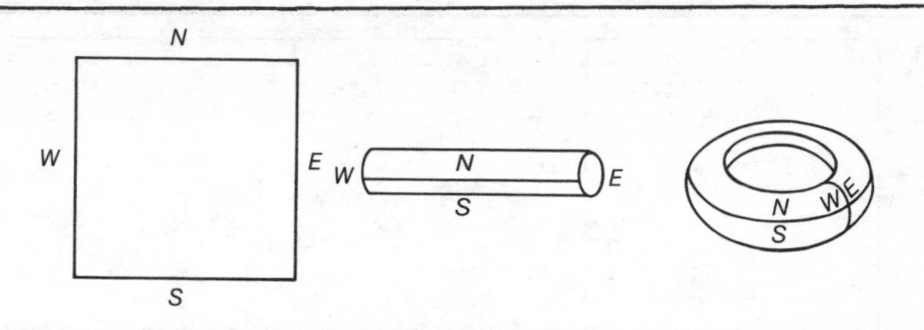

Figure 3.3 Creation of a Torus from a Planar Surface

This procedure, illustrated in Figure 3.4b, is equivalent to ignoring all points in the pattern that are closer to the study area boundary than they are to any point in the pattern. This *disregard* strategy has the effect of reducing the number of distances that can be measured. This may be particularly awkward if N is already small or if we are measuring higher-order neighbor distances. It also leads to bias because it favors the retention of small d_i's.

A third strategy for overcoming the boundary problem, illustrated in Figure 3.4c, is to delimit the study area as a smaller part of the entire point pattern. The area of the point pattern outside of the study area is known as the "*buffer zone.*" Nearest neighbor distances are only measured for those points within the study area even though these may be distances to points in the buffer zone. This procedure also has the effect of reducing the number of measurements.

A fourth solution to the boundary problem can be used if toroidal mapping is not possible or if the disregard and buffer zone strategies reduce the number of distances prohibitively. This strategy is to add a "correction factor" to equations 13 and 15 to account for the boundary effects. Donnelly (1978) has shown that when N is greater than 7 and the study area is not highly irregular, the value of $E(d_i)$ is approximated by

$$E(d_i) = 0.5 \sqrt{(A/N)} + (0.0514 + 0.041 / \sqrt{N})\, B/N \quad [16]$$

and

$$\text{var}\,(\bar{d}) = 0.070\, A/N^2 + 0.037\, B \sqrt{(A/N^5)} \quad [17]$$

where B is the length of the perimeter of the study area. Since equations 16 and 17 were obtained by examining simulated point patterns in study

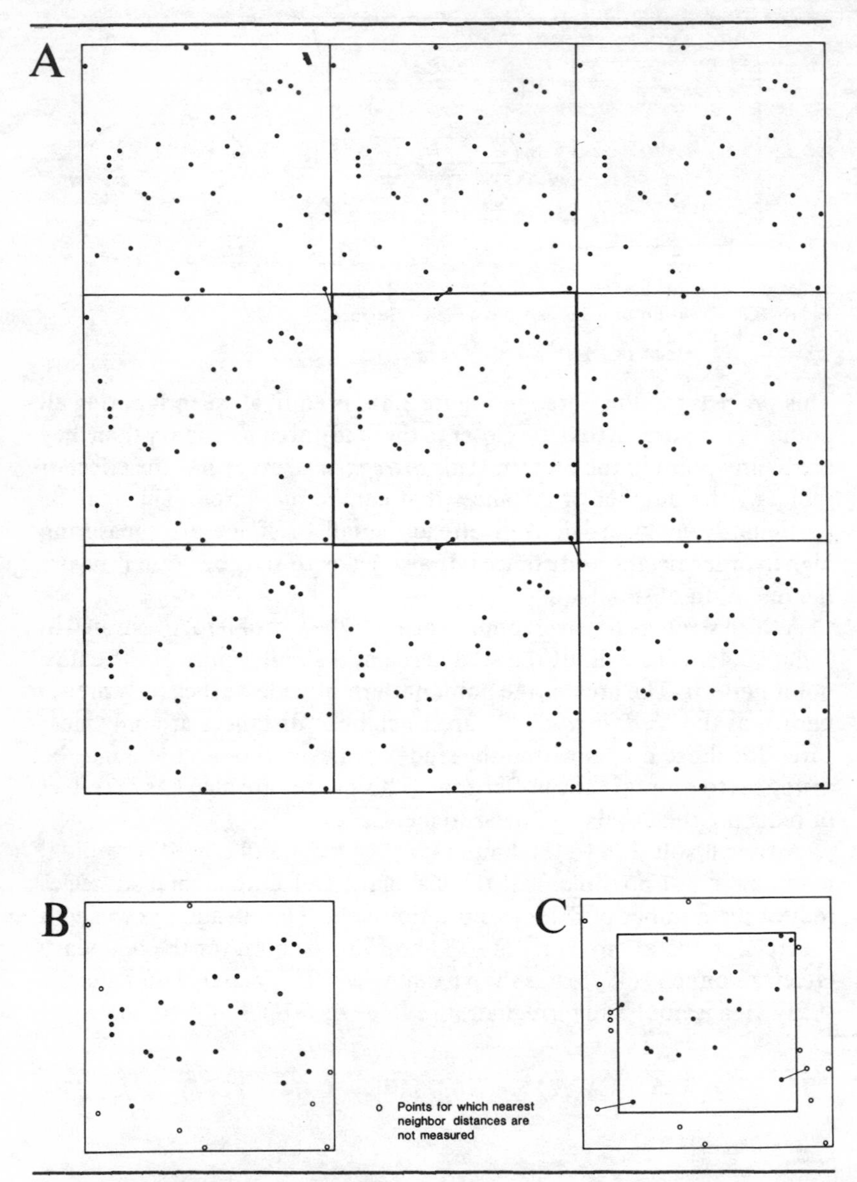

Figure 3.4 Methods of Dealing with the Effects of Study Area Boundaries: (a) Toroidal Mapping, (b) Disregard, and (c) Buffer Zone

areas of various shapes including circles, ellipses, squares, and rectangles, they should not be used for irregularly shaped study areas. Examples of such shapes where this strategy may not be used are shown in Figure 3.5. The calculated values of $E(d_i)$ and $var(\bar{d})$ from equations 16 and 17, respectively, are used in equation 14 to calculate the z statistic.

The following example illustrates the use of Donnelly's solution to the boundary problem. In Figure 3.6 the locations are shown of branches of the Royal Bank that had computerized teller facilities in metropolitan Toronto (excluding the eastern borough of Scarborough) in September 1982. In this example the shape of the study area prohibits toroidal mapping, whereas the limited size of the pattern ($N = 25$) together with the peripheral location of several of the points makes either the disregard or buffer zone methods undesirable; the disregard technique reduces the number of points to 16. The value of $\bar{d}$ is found to be 1.89 kms., B is 92.47 kms, and A is 453.26 sq. kms. Using equation 16 we get

$$E\,(d_i) = 0.5\,\sqrt{(453.26/25)} + (0.0514 + 0.041\ /\ \sqrt{25})\ 92.47/25$$
$$= 2.35 \text{ kms.}$$

and using equation 17

$$var\,(\bar{d}) = 0.070\ (453.26)\ /\ (25)^2 + 0.037\ (92.47)\ \sqrt{(453.26/25^5)}$$
$$= 0.0741$$

Substituting these calculated values of $E(d_i)$ and $var(\bar{d})$ in equation 14 we get

$$z = \frac{1.89 - 2.35}{\sqrt{0.0741}}$$

$$= -1.69.$$

This value of z does not exceed the value of z from tables of the normal distribution for $\alpha = 0.05$ ($z = 1.96$). Therefore, the H_0 is accepted. This acceptance suggests that the location of the bank branches is not unlike a CSR pattern. Perhaps this result is the outcome of the operation of conflicting forces of dispersion caused by attempting to spread out the branches over the metropolitan area to increase consumer access, versus the inhomogeneity of the demand for such services resulting from a

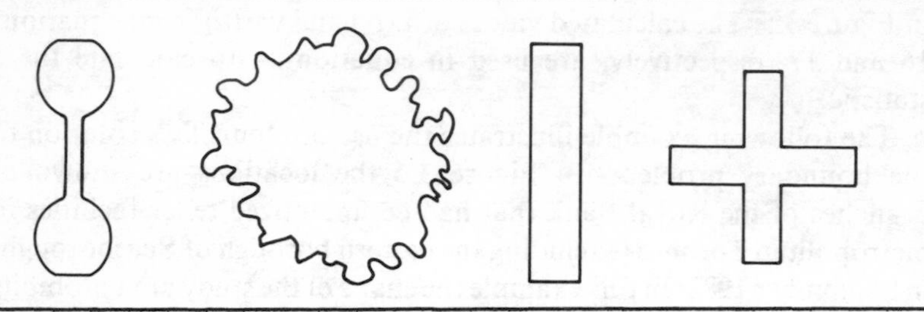

Figure 3.5 Examples of Study Area Shapes for Which Donnelly's Correction Formulas Are Inappropriate

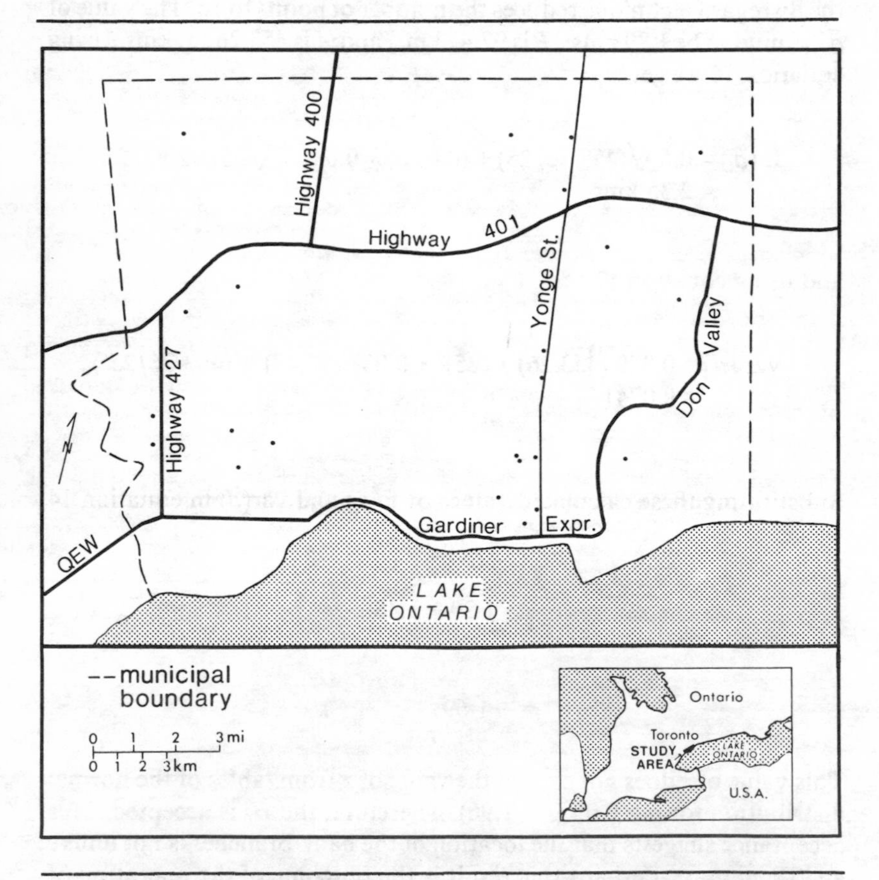

Figure 3.6 Location of Royal Bank Branches with Computer Facilities in Metropolitan Toronto (Excluding the Municipality of Scarborough) in September 1982

higher demand for banking services in the main financial and retailing districts of the metropolitan area and a lower demand in manufacturing and suburban residential districts.

Examples of the use of nearest neighbor analysis in geography to study patterns in two dimensions include the work of King (1961, 1962), Dacey (1962, 1963), and Birch (1967) on urban settlement locations; Getis (1964), Clark (1969), and Sherwood (1970) on the locations of intra-urban facilities; Smalley and Unwin (1968), Jauhiainen (1975), and Rose and Letzer (1975) on drumlins; Williams (1972b) and Day (1976) on dolines; and Morgan (1970) on river basin outlets.

3.4 Nearest Neighbor Analysis in One Dimension

So far in this book we have considered patterns where the points represent locations in a two-dimensional study area. However, geographers often encounter situations where phenomena can be represented as points along a line. Examples include cities along a highway, retail stores along a street, effluent discharge outlets along a river, and volcanoes along a fault line.

As with points in two dimensions we can measure the distance, d_i, to the nearest neighbor of each point, *i*, along the line. Lines, like areas, can be bounded. However, lines are more simple than areas because each line is limited to a maximum of two boundaries: the ends of the line.

Selkirk and Neave (1984) consider two situations for the relationship of the points in the pattern to the ends of the line: points located at both ends of the line and no points located at either end of the line.

When points are located at both ends of the line, the expected value of the average nearest neighbor distance, $E(d_i)$ for a CSR pattern is

$$E(d_i) = [W(N+2)] \; / \; [2N(N-1)] \qquad [18]$$

and the variance, $var(\bar{d})$, is

$$var(\bar{d}) = [W^2(2N^2 + 7N - 36)] \; / \; [12N^3 \; (N-1)^2] \qquad [19]$$

where W is the length of the line. These values can be used in equation 14 to calculate a z statistic that has an approximately normal distribution. When N is small, say $N \leq 20$, equation 14 should not be used. Instead, the appropriate value can be found in Table 1 of Selkirk and Neave (1984).

In the case where no points are located at the ends of the line the appropriate values of $E(d_i)$ and $\mathrm{var}(\bar{d})$ are

$$E(d_i) = [W(N+2)] / [2N(N+1)] \qquad [20]$$

and

$$\mathrm{var}(\bar{d}) = [W^2(2N^2 + 17N + 12)] / [12N^2(N+1)^2(N+2)]. \qquad [21]$$

When $N > 20$ these values can also be used in equation 14 to yield a z statistic that has an approximately normal distribution. If $N \leq 20$, the exact tables published in Selkirk and Neave (1984) should be consulted.

As an example of a pattern of points along a line, consider Figure 3.7, which shows the locations of interchanges along Highway 401 (the main east-west provincial highway in Ontario, Canada) as it passes through metropolitan Toronto. There are 24 interchanges and the distance between the eastern and western boundaries of metropolitan Toronto is 44.4 kms. The value of $\bar{d}$ is 1.37 kms.

Since the ends of the line are demarcated by the eastern and western boundaries of metropolitan Toronto, we use equations 20 and 21.

From eqation 20, $E(d_i)$ is 0.96 kms and from equation 21 $\mathrm{var}(\bar{d}) = 0.0276$. Thus

$$z = (1.37 - 0.96) / \sqrt{0.0276}$$

$$= 2.47.$$

Since this value of z exceeds the value of 1.96 from tables of the normal distribution for $\alpha = 0.05$, we reject the H_0. Further, since $\bar{d} > E(d_i)$ there is evidence that the interchanges are regularly spaced.

In those situations where the study area is a line, the nearest neighbor distance for any point not located at one of the end points must be the distance to either the preceeding point or the succeeding point encountered on the line. Thus the nearest neighbor distances are part of the set of all interpoint distances on the line. Several tests make use of this more extensive information. One such test uses a statistic, S, suggested by Durbin (1965).

In Durbin's test the interpoint distances are first converted to proportions of the sum of the interpoint distances. The resulting scaled values, g_i, are then ranked from the smallest, g_1, to the largest, g_n; where

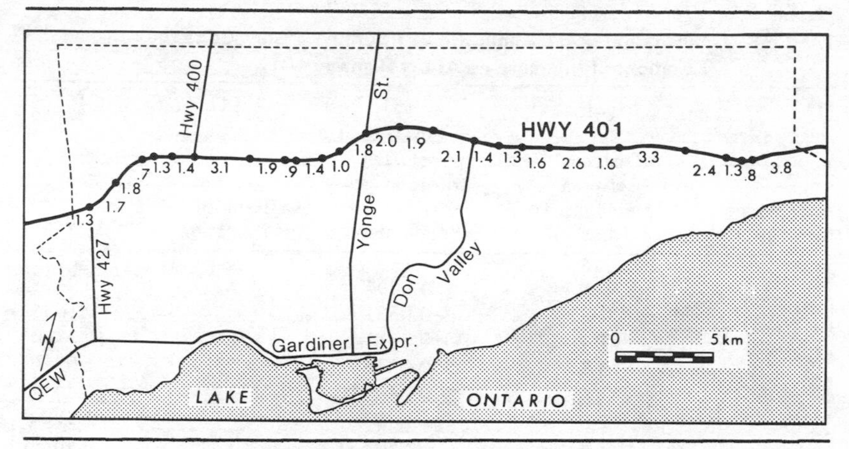

Figure 3.7 Locations of Interchanges Along Highway 401 in Metropolitan Toronto

n is the number of interpoint distances. These values are used to calculate the statistic, S, where

$$S = 2n - 2 \sum_{i=1}^{n} i g_i. \quad [22]$$

The expected value of this statistic for a CSR pattern is given by

$$E(S) = (n-1)/2 \quad [23]$$

with variance

$$\text{var}(S) = (n-1) / 12. \quad [24]$$

The observed and expected values may be compared using a normally distributed statistic, z of the form

$$z = [S - E(S)] / \sqrt{\text{var}(S)}. \quad [25]$$

Durbin's S statistic can be illustrated using information from the previous example of the location of interchanges along Highway 401. The 23 interpoint distances between the highway interchanges are given in column 2 of Table 3.2. Column 3 of Table 3.2 shows these interpoint distances as proportions of the sum of interpoint distances of 39.3 kms.

TABLE 3.2 Illustration of the Application of Durbin's *S* Statistic to the Locations of Interchanges Along Highway 401

[1] Interchange Number	[2] Distance Between Interchanges (kms)	[3] Distance Between Interchanges as Proportions	[4] Rank of Proportion	[5] [3] × [4]
63				
	0.8	0.0204	2	0.0408
62				
	1.3	0.0331	5	0.1655
64				
	2.4	0.0611	20	1.2220
61A				
	3.3	0.0840	23	1.9320
60				
	1.6	0.0407	11	0.4477
59A				
	2.6	0.0662	21	1.3902
59				
	1.5	0.0407	12	0.4884
58				
	1.3	0.0331	6	0.1986
57				
	1.4	0.0356	8	0.2848
56				
	2.1	0.0534	19	1.0146
55				
	1.9	0.0483	16	0.7728
54				
	2.0	0.0509	18	0.9162
53				
	1.8	0.0458	14	0.6412
52				
	1.0	0.0254	4	0.1016
51A				
	1.4	0.0356	9	0.3204
51				
	0.9	0.0229	3	0.0687
50A				
	1.9	0.0483	17	0.8211
50				
	3.1	0.0789	22	1.7358
49				
	1.4	0.0356	10	0.3560
48				
	1.3	0.0331	7	0.2317
47				
	0.7	0.0178	1	0.0178
46A				
	1.8	0.0458	15	0.6870
46				
	1.7	0.0433	13	0.5629
45				
Total	39.3	1.0000	–	14.4178

These proportions are then ranked from the smallest (0.0178) to the largest (0.0840). The ranking of each proportion is given in column 4 of Table 3.2. Inserting the value from column 5 into equation 22 gives

$$S = 2(23) - 2(14.4178)$$

$$= 17.1644.$$

Equations 23 and 24 yield $E(S) = 11$ and $\text{var}(S) = 1.8333$, respectively. Substituting these values into equation 25 yields $z = 4.55$. Since this calculated value of z is both positive and larger than the value of $z = 1.96$

($\alpha = 0.05$) obtained from the tables of the normal distribution, the H_0 is rejected in favor of one that indicates regularity in the point pattern, confirming the result of the nearest neighbor test applied previously to the data.

3.5 Refined Nearest Neighbor Analysis

The distance techniques examined above involve the calculation of a single summary statistic. In calculating the statistic it has been necessary to measure a number of individual distances. The conversion of these sets of individual distances into a single statistic represents a loss of information. This situation has prompted several researchers (Cowie 1967; Campbell and Clark 1971; Roder 1974, 1975; Diggle 1979a) to suggest using tests that retain the original information.

The most common of such tests has been called refined nearest neighbor analysis (Diggle 1979a). It involves comparing the complete distribution function of the observed nearest neighbor distances, $F(d_i)$, with the distribution function of expected nearest neighbor distances for CSR, $P(d_i)$.

$F(d_i)$ is obtained by taking the nearest neighbor distances, d_i, and ranking them from the smallest to the largest. Once we have done this we are able to determine what proportion $F(d_i \leq r)$, of the nearest neighbor distances are less than or equal to some chosen distance, r. Usually, the values of r are chosen so that they correspond with values of d_i. Pielou (1969: 111-112) shows that the corresponding proportion of expected nearest neighbor distances less than or equal to r for an unbounded CSR pattern $P(d_i \leq r)$ is given by equation 26.

$$P(r) = 1 - e^{-\pi r^2 \lambda} \qquad [26]$$

where:

e is the mathematical constant 2.718283
π is the mathematical constant 3.141593
r is the specified distance
λ is estimated from N/A

Diggle (1981) has suggested that $F(r)$ and $P(r)$ can be compared using a statistic, d_r, of the form

$$d_r = \max | F(r) - P(r) | \qquad [27]$$

where max | • | means the largest absolute value obtained for corresponding values of r. Since the observed nearest neighbor distances are not mutually independent, Diggle (1981: 26) suggests that in order to evaluate the significance of d_r we must use a Monte Carlo test procedure. This involves generating a set (usually 99) of CSR patterns each with the same number of points as the empirical pattern located in a study area identical to that of the empirical pattern (Diggle 1979a). d_r is calculated for each of these simulated patterns. We can then examine where the value of d_r for the empirical pattern falls within the entire set of 100 values (99 from the simulated patterns plus one from the empirical pattern) thus giving us an indication of the likelihood of the real pattern occurring under the conditions of the random process described in section 1.3. Should the value of d_r for the empirical pattern be among the five largest of the 100 values of d_r, the H_0 of CSR can be rejected for α = 0.05. Further, Diggle (1979a) suggests that if for d_r, $F(r) > P(r)$, then a clustered pattern is indicated, whereas $F(r) < P(r)$ indicates a regular pattern of points.

To illustrate the use of refined nearest neighbor analysis, consider Figure 3.8, which shows the locations of 74 central places in south-central England that provided bus services to surrounding rural areas in the 1950s (Green 1955). We begin by measuring the nearest neighbor distance, d_i, for each point, i, in the pattern. The individual d_i values are then ranked and converted to proportions. To find the proportion of points in the pattern for which d_i is less than or equal to, for example, r = 1.95 miles, we count the number of points for which $d_i \leq r$ and divide this number by the total number of points, N. However, since equation 26 relates to an unbounded CSR pattern, we must make allowance for the impact of edge effects on the pattern under study. To do this we measure the distance, u_i, of each point, i, from the study area boundary. We then subtract from N the number of points for which $d_i > r$ and $u_i < r$. Thus as r changes, the number of points subtracted from N may change. For the pattern in Figure 3.8, for r = 1.95 miles, there are 2 points for which $d_i \leq r$ (see column 2 of Table 3.3) and 11 points for which $d_i > r$ and $u_i < r$ (see column 3 of Table 3.3) so that the observed proportion, $F(r)$, shown in column 4 of Table 3.3 is given by $2/(74 - 11) = 0.0317$. For r = 2.66 miles, $d_i \leq r$ for 4 points and $u_i < r < d_i$ for 12 points so that $F(r) = 4/(74 - 12)$ = 0.0645. The remaining values of $F(r)$, given in column 4 of Table 3.3, are found in a similar way.

We use equation 26 to calculate the values of $P(r)$. In this example N = 74 and A = 7259.71 sq. miles, so that λ is estimated from $N/A = 0.0102$. Thus

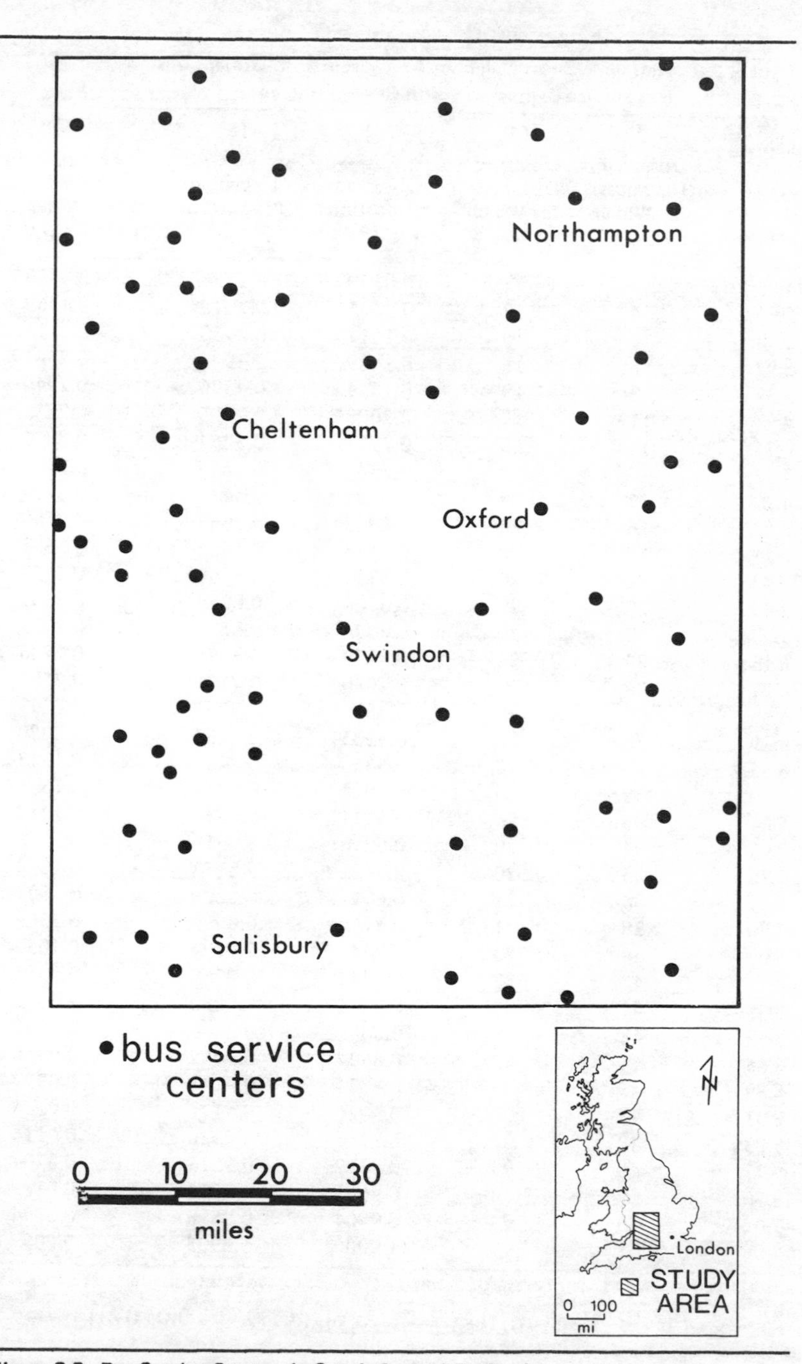

Figure 3.8 Bus Service Centers in South-Central England

TABLE 3.3 Refined Nearest Neighbor Analysis in Two-Dimensions: Bus Service Centers in South-Central England

[1] r (miles)	[2] Observed Distances for Which $d_i \leq r$	[3] Observed Distances for Which $u_i < r < d_i$	[4] Observed Proportion $F(r)$	[5] Expected Proportion $P(r)$	[6] $\lvert F(r) - P(r) \rvert$
1.95	2	11	0.0317	0.1135	0.0818
2.66	4	12	0.0645	0.2015	0.1369
3.44	6	15	0.1017	0.3133	0.2116
3.97	8	18	0.1429	0.3929	0.2501
4.39	9	19	0.1636	0.4576	0.2940
4.63	11	19	0.2000	0.4935	0.2935
4.71	12	19	0.2182	0.5054	0.2873
4.81	14	19	0.2345	0.5195	0.2650
4.97	15	19	0.2727	0.5436	0.2709
5.13	16	20	0.2963	0.5662	0.2699
5.26	17	20	0.3148	0.5836	0.2688
5.42	18	20	0.3333	0.6066	0.2732
5.63	20	20	0.3704	0.6344	0.2641
6.00	22	20	0.4074	0.6803	0.2729
6.16	23	20	0.4259	0.6995	0.2736
6.27	25	20	0.4630	0.7131	0.2501
6.30	26	20	0.4815	0.7164	0.2349
6.39	28	20	0.5185	0.7265	0.2080
6.53	30	20	0.5556	0.7409	0.1853
6.59	32	20	0.5926	0.7481	0.1555
6.64	33	20	0.6111	0.7528	0.1417
7.03	34	20	0.6296	0.7912	0.1616
7.30	35	20	0.6481	0.8153	0.1672
7.47	36	21	0.6792	0.8293	0.1501
7.78	38	21	0.7170	0.8532	0.1362
8.03	39	23	0.7647	0.8704	0.1057
8.11	40	23	0.7843	0.8759	0.0916
8.31	42	23	0.8235	0.8880	0.0645
8.59	43	23	0.8431	0.9036	0.0605
8.68	44	23	0.8627	0.9082	0.0455
8.89	45	23	0.8824	0.9143	0.0319
8.91	46	23	0.9020	0.9174	0.0154
9.23	47	23	0.9216	0.9330	0.0114
9.77	48	23	0.9412	0.9516	0.0104
10.70	49	23	0.9607	0.9736	0.0129
12.02	50	23	0.9804	0.9898	0.0094
12.09	51	23	1.0000	0.9903	0.0097

$$P(r \leq 1.95 \text{ miles}) = 1\text{-}\exp[-\pi(1.95)^2(0.0102)]$$

$$= 0.1135,$$

$$P(r \leq 2.66 \text{ miles}) = 1\text{-exp}[-\pi(2.66)^2(0.0102)]$$

$$= 0.2015.$$

The complete set of $P(r)$ values obtained in this way is given in column 5 of Table 3.3.

The values of $F(r)$ and $P(r)$ in columns 4 and 5, respectively, are then compared for corresponding values of r (see column 6 of Table 3.3) and the largest of the values in column 6 of Table 3.3 is equal to d_r as defined in equation 27. In this case $d_r = 0.2940$ at $r \leq 4.39$. Further, since $F(r) < P(r)$ at this value of r, a regular pattern is indicated.

In order to determine if the value of $d_r = 0.2940$ is sufficient to reject an H_0 of a CSR pattern for the bus service centers, it is necessary to perform the Monte Carlo test procedure described earlier. This involves generating 99 patterns according to the two conditions in described section 1.3 that produce a CSR pattern. In each of the 99 simulations we locate the same number of points as in the empirical pattern—namely, 74—over a study area of the same size and shape as the empirical study area shown in Figure 3.8. For each of the 99 CSR patterns we calculate d_r. As Table 3.4 shows, of the 99 runs generated, only one produced a d_r value greater than that for the empirical pattern and so we conclude that the H_0 of CSR can be rejected in favor of a more regular pattern.

A similar procedure may be applied when the study area is a line (Roder 1974). However, in this case the distribution function of expected nearest neighbor distances for an unbounded CSR pattern is given by

$$P(r) = 1 - e^{-2r\lambda} \quad [28]$$

where:

λ is the density of points per unit length.

To illustrate this procedure, consider Figure 3.9, which shows the locations of interchanges on the western part of provincial Highway 401 in southern Ontario, between its beginning at the U.S. border at Windsor and the western boundary of metropolitan Toronto.

We begin by identifying the nearest neighbor distances. These are ranked from the smallest to the largest (see column 1 of Table 3.5). To allow for edge effects in calculating $F(r)$ we ignore those points for which $d_i > r$ and $u_i < r$. In this way the interchange closest to the border of

TABLE 3.4 Cumulative Distribution of d_r for 99 Monte Carlo Simulations of Random Patterns in Two Dimensions

Value of d_r	Frequency $\geq d_r$
0.06	99
0.08	95
0.10	81
0.12	62
0.14	46
0.16	23
0.18	10
0.20	5
0.22	5
0.24	1
0.26	1
0.28	1
0.30	1

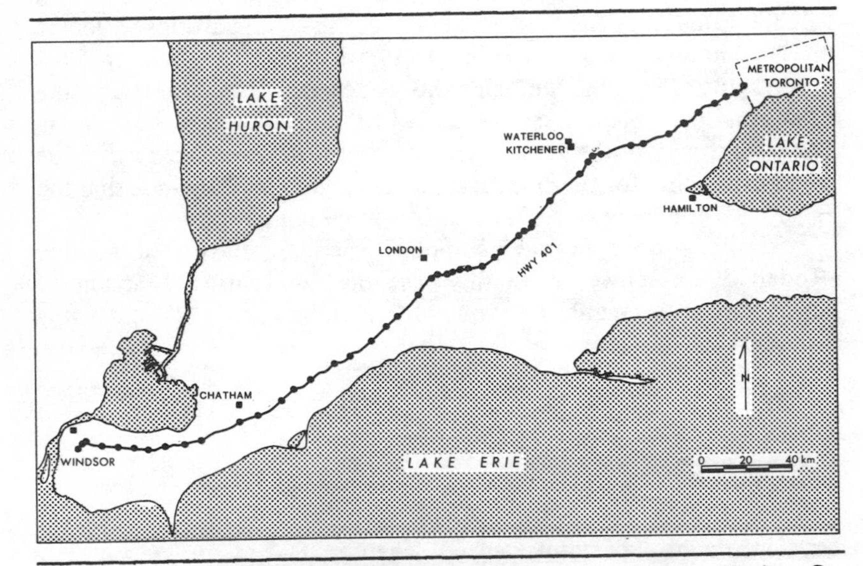

Figure 3.9 Locations of Interchanges Along Highway 401 Between Windsor, Ontario, and the Western Boundary of Metropolitan Toronto

metropolitan Toronto is removed from the analysis. Thus, for example, since there are 47 interchanges, $F(r \leq 0.8 \text{ kms}) = 2/(47-1) = 0.0435$. This and the remaining values of $F(r)$ are shown in column 3 of Table 3.5.

Equation 28 is used to obtain the values of $P(r)$ for the corresponding values of r. In equation 28, λ can be estimated by using the ratio of the number of points to the line length in the empirical pattern. Since $N = 47$

TABLE 3.5 **Refined Nearest Neighbor Analysis in One Dimension: Interchanges Along Highway 401, Ontario**

[1] r (kms)	[2] Observed Number $d_i \leqslant r$	[3] Observed Proportion $F(r)$	[4] Expected Proportion $P(r)$	[5] $\|F(r) - P(r)\|$
0.8	2/46	0.0435	0.1958	0.1524
1.6	4/46	0.0870	0.3533	0.2664
1.9	6/46	0.1304	0.4041	0.2737
2.3	8/46	0.1739	0.4656	0.2917
2.7	10/46	0.2174	0.5208	0.3034
2.9	12/46	0.2609	0.5262	0.2853
3.4	14/46	0.3043	0.6040	0.2997
3.9	16/46	0.3478	0.6544	0.3066
4.3	18/46	0.3913	0.6901	0.2988
4.5	21/46	0.4565	0.7066	0.2501
5.6	22/46	0.4783	0.7826	0.3043
5.8	23/46	0.5000	0.7941	0.3941
6.2	25/46	0.5435	0.8153	0.2718
6.3	26/46	0.5652	0.8203	0.2551
6.8	28/46	0.6086	0.8432	0.2345
6.9	30/46	0.6522	0.8474	0.1952
7.1	32/46	0.6957	0.8555	0.1598
7.3	33/46	0.7174	0.8632	0.1458
7.4	35/46	0.7609	0.8668	0.1059
7.9	36/46	0.7826	0.8838	0.1012
8.2	41/46	0.8913	0.8929	0.0016
8.6	43/46	0.9348	0.9040	0.0308
9.2	44/46	0.9565	0.9185	0.0380
9.5	45/46	0.9783	0.9249	0.0534
12.2	46/46	1.0000	0.9640	0.0360

and $W = 345$ kms, $\lambda = 0.1362$. Thus $P(r \leq 0.8 \text{ kms}) = 1 - \exp[-2(0.8)(0.1362)] = 0.1958$. This and the other values of $P(r)$ obtained similarly are shown in column 4 of Table 3.5.

As with refined nearest neighbor analysis in two dimensions the corresponding values of $F(r)$ and $P(r)$ are used to define the d_r statistic in equation 27. In this case $d_r = 0.3066$ for $r \leq 3.9$. The significance of d_r is tested using the Monte Carlo approach described earlier. This involves generating 99 CSR patterns each with the same number of points as the empirical pattern located along a line of the same length as the empirical one and calculating d_r for each of the simulated patterns. Since none of the d_r values obtained from the simulated patterns exceeds 0.30, (see Table 3.6), the H_0 of CSR is clearly rejected at $\alpha = 0.05$. Further, since at

TABLE 3.6 **Cumulative Distribution of d_r for 99 Monte Carlo Simulations of Random Patterns in One Dimension**

Value of d_r	Frequency $\geqslant d_r$
0.06	99
0.08	94
0.10	79
0.12	63
0.14	48
0.16	28
0.18	17
0.20	10
0.22	6
0.24	3
0.26	1
0.28	1
0.30	0

$r \leq 3.9$, $F(r) < P(r)$, regularity in the location of interchanges along Highway 401 is indicated.

In those instances of a linear study area in which interpoint distances have been calculated, as was the case with Durbin's S statistic in section 3.4, a similar approach may be used except that the expected values in a CSR pattern are given by

$$P(r) = 1\text{-}e^{r\lambda}. \qquad [29]$$

3.6 Second-Order Analysis

Another technique used to examine the dispersion characteristics of a point pattern is second-order analysis. This technique requires as data the distance measurements between all combinations of pairs of points. In its descriptive form it is the study of interevent distances, where the events are mapped points. In its theoretical form, it is called second-order analysis to indicate that the focus is on the variance, or second moment, of interevent distances.

The study of interevent distances has a number of advantages over the more traditional techniques of analysis. First, the data set yielded by even a modest number of points allows for a detailed view of pattern spacing characteristics. More information about pattern is potentially available from this analysis than from any other existing technique. Second, the CSR model available for the interevent distances can be used as the basis for tests for statistical significance (second-order analysis). Third, because a statistically defensible boundary correction technique has been developed for second-order studies, arbitrary border correction schemes are not necessary. Finally, a convenient method exists for the study of various distance subdivisions, or distance zones.

In the following pages we describe in detail the analysis of the population pattern of McHenry County, Illinois (see Figure 3.10). The location of 9 centers of population were identified by use of census data; each location is noted by a number. The analysis of interevent distances is based on the number of pairs of points, *i, j,* identified within a given distance, *d,* of each of the *i* points. Imagine centering a circle of radius *d* on each point. Each of the points covered by the circle is paired with the center point of that circle, and it is this number of pairs that forms our data (see Figure 3.11). As the size of the circle is increased uniformly around each point, we note the increasing number of pairs of points contained within the circles. In the usual analysis, the value of *d* is increased from 0 to a distance beyond which statistical bias (discussed later) is evident.

The analysis of interevent distances depends on the expected number of pairs of points in a Poisson process (the CSR model). Our task is to compare the number of observed pairs with the expectation at all distances, taking into consideration the density of points, the borders, and the size of the sample.

The CSR formula given by Ripley (1981: 159-160) is

$$L(d) = [(A \; S \; k \; (i,j)) \; / \; \pi N \; (N-1)]^{1/2} \qquad [30]$$

where $L(d)$ is a linear expression of the expected number of events (points) within distance d of all i events (points), and where

$$S \text{ denotes } \sum_{i=1}^{N} \sum_{\substack{j=1 \\ i \neq j}}^{N}$$

and $S\,k(i,j)$ is defined as the number of j points within distance d of all i points, and A is the size of the study area. Note that the summation sign calls for the enumeration of all pairs within d except when i and j are the same; thus a center point is not paired with itself. The square root sign and π serve to make $L(d)$ a linear function of d. In fact, the expectation of $L(d)$ is d.

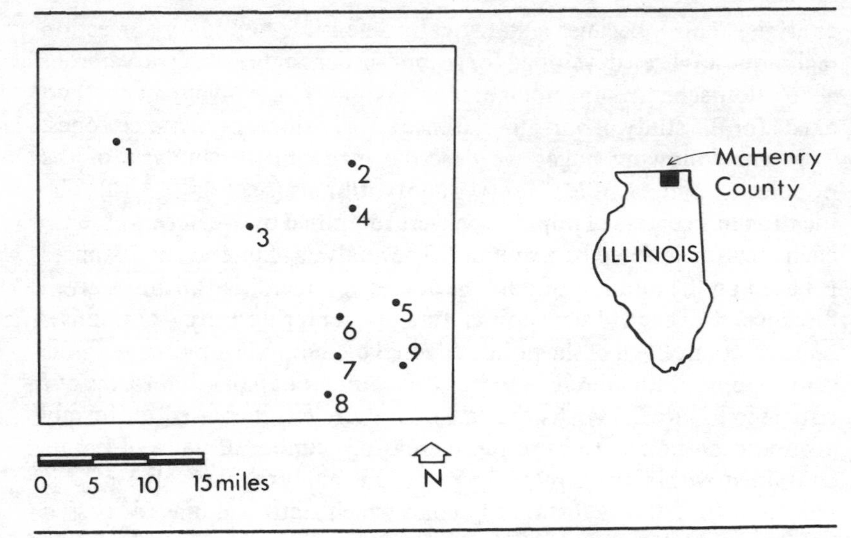

Figure 3.10 Location of the 9 Principal Centers of Population in McHenry County, Illinois, in 1980

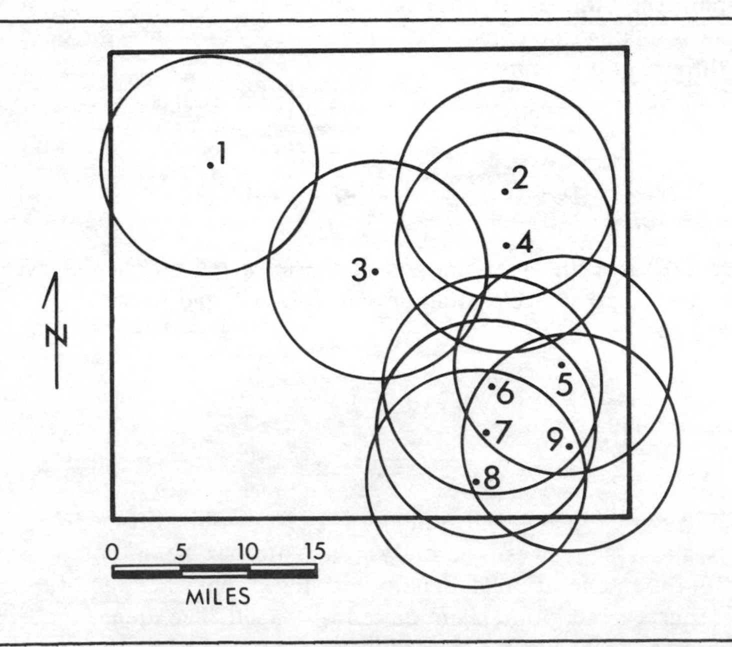

Figure 3.11 Circles of Radius d = 10 Miles Centered on Points 1 to 9 of McHenry County (24 Pairs of Points are Encircled)

If no border corrections are employed, beyond a certain point d the number of additional pairs per unit distance decreases. This bias is of no interest if the goal is simply to describe the pattern of points. In order to test for CSR, however, one must assume that the point process is continuous beyond the boundaries of the study area. Before the edge correction is given, refer to Table 3.7. This gives the matrix of all $N(N-1)$ or 72 possible distances for the 9 points of McHenry County.

The data in Table 3.7 are given in miles. Note that the table is symmetrical; that is, the distances are the same in each direction. For example, the distance from point 1 to point 5 equals the distance from point 5 to point 1. Instead of using straight line distances, one might have used transportation distances or travel time, where such factors as one-way streets could make the distance matrix asymmetrical. In that case, the number of distinct distance measures would be $N(N-1)$ instead of $N(N-1)/2$.

Note, too, that the number of ordered pairs of points make up the data base. For illustration, when i is set at 1 and j at 5 and d is greater or equal to the distance from point 1 to point 5, the combination 1,5 counts as one pair. When i is set at 5 and j is 1, the combination 5,1 counts as another pair within d.

Distances are calculated using the Pythagorean theorem. First, the coordinates of each point $X(i)$, $Y(i)$ are identified. The distance d separating points i and j is

$$d = \surd([X(i) - X(j)]^2 + [Y(i) - Y(j)]^2) \qquad [31]$$

In Table 3.8, columns 1, 2, and 3 identify the McHenry County point pairs by distance. In Figure 3.12 the number of pairs are graphed by distance, revealing, as expected, that the number of pairs increases at a decreasing rate toward the right-hand side of the curve.

The assumption of continuity beyond the boundaries requires that any point beyond the boundary must be counted within the circle centered on a point in the study area. We assume that the pattern of points outside of the study area is like the nearby pattern within the study area. Mathematically, this is accomplished by weighting pairs of points within d whose distance apart (i to j) is greater than the distance that point i is to the border. The weight will increase the value of the pair to something greater than one, so as to take into account the "missing" points from outside of the study area.

The procedure for weighting can be made clear if we focus our attention on points 7 and 8 of Figure 3.11. Figure 3.13a shows the area

TABLE 3.7 Straight-Line Distances in Miles Between all 9 Centers of Population in McHenry County, Illinois

Population Center	1	2	3	4	5	6	7	8	9
1	–	22.09	14.84	22.94	30.02	27.02	28.64	30.92	34.12
2	22.09	–	11.24	4.50	13.60	15.03	18.06	22.14	20.01
3	14.84	11.24	–	9.62	15.21	12.38	14.42	17.46	19.45
4	22.94	4.50	9.62	–	9.39	10.55	13.58	17.68	15.66
5	30.02	13.60	15.21	9.39	–	5.39	7.43	11.10	6.52
6	27.02	15.03	12.38	10.55	5.39	–	3.04	7.16	7.11
7	28.64	18.06	14.42	13.58	7.43	3.04	–	4.16	6.18
8	30.92	22.14	17.46	17.68	11.10	7.16	4.12	–	7.43
9	34.12	20.01	19.45	15.66	6.52	7.11	6.18	7.43	–

TABLE 3.8 Second-Order Data for the Population Centers of McHenry County, Illinois

[1] Pair i,j	[2] Pair Number	[3] d	[4] K (i,j)	[5] S K (i,j)	[6] L (d)	[7] L (d)−d
6,7	1	3.04	1.00			
7,6	2	3.04	1.00	2.00	3.38	0.34
7,8	3	4.11	1.00			
8,7	4	4.12	1.47	4.47	5.05	0.83
2,4	5	4.50	1.00			
4,2	6	4.50	1.00	6.47	6.07	1.57
5,6	7	5.39	1.07			
6,5	8	5.39	1.00	8.54	6.98	1.59
7,9	9	6.18	1.17			
9,7	10	6.18	1.98	11.68	8.16	1.99
5,9	11	6.52	1.25			
9,5	12	6.52	2.05	14.98	9.24	2.72
6,9	13	7.11	1.00			
9,6	14	7.11	2.15	18.13	10.17	3.06
6,8	15	7.16	1.00			
8,6	16	7.16	1.67	20.80	10.89	3.76
5,7	17	7.43	1.30			
7,5	18	7.43	1.33			
8,9	19	7.43	2.20			
9,8	20	7.43	1.68	27.31	12.48	5.05
4,5	21	9.39	1.35			
5,4	22	9.39	1.45	30.11	13.10	3.71
3,4	23	9.62	1.00			
4,3	24	9.62	1.37	32.48	13.61	3.99
4,6	25	10.55	1.41			
6,4	26	10.55	1.65	36.54	14.43	3.88
5,8	27	11.10	1.93			
8,5	28	11.10	1.80	40.27	15.15	4.05
2,3	29	11.24	1.58			
3,2	30	11.24	1.00	42.85	15.63	4.39
3,6	31	12.38	1.00			
6,3	32	12.38	1.90	45.74	16.15	3.77
4,7	33	13.58	2.08			
7,4	34	13.58	2.18	50.00	16.89	3.31
2,5	35	13.60	1.91			
5,2	36	13.60	2.22	54.13	17.57	3.97
3,7	37	14.42	1.00			
7,3	38	14.42	2.25	57.39	18.09	3.67
1,3	39	14.84	2.27			
3,1	40	14.84	1.00	60.66	18.60	3.76
2,6	41	15.03	2.04			
6,2	42	15.03	2.14	64.84	19.23	4.20
3,5	43	15.21	1.00			

Continued

TABLE 3.8 Continued

[1] Pair *i,j*	[2] Pair Number	[3] *d*	[4] *K (i,j)*	[5] *S K (i,j)*	[6] *L (d)*	[7] *L (d)−d*
5,3	44	15.21	2.36	68.20	19.72	4.51
4,9	45	15.66	2.26			
9,4	46	15.66	2.93	73.39	20.46	4.80
3,8	47	17.46	1.11			
8,3	48	17.46	2.63	77.13	20.97	3.51
4,8	49	17.68	2.40			
8,4	50	17.68	2.64	82.17	21.65	3.97
2,7	51	18.06	2.26			
7,2	52	18.06	2.51	86.94	22.26	4.20
3,9	53	19.45	1.21			
9,3	54	19.45	3.09	91.24	22.81	3.36
2,9	55	20.01	2.37			
9,2	56	20.01	3.11	96.72	23.48	3.47
1,2	57	22.09	2.68			
2,1	58	22.09	2.47	101.87	24.10	2.01
2,8	59	22.14	2.48			
8,2	60	22.14	2.86	107.21	24.72	2.58
1,4	61	22.94	2.72			
4,1	62	22.94	2.66	112.59	25.34	2.40
1,6	63	27.02				
6,1	64	27.02				
1,7	65	28.64				
7,1	66	28.64				
1,5	67	30.02		Region of bias		
5,1	68	30.02				
1,8	69	30.92				
8,1	70	30.92				
1,9	71	34.12				
9,1	72	34.12				

around point 8 in greater detail. If the boundary is a straight line, as it is in this case, the following formula given by Getis (1984) can be used to find the contribution of the pair 8,7 to the total number of pairs within *d*:

$$k(i,j) = \{1 - [\cos^{-1} (e/d) / \pi]\}^{-1} \qquad [32]$$

where *e* is the distance to the nearest edge. This formula describes the inverse of the ratio of the fraction of the circumference of the circle contained within the study area to the whole circumference. The longer the circumference of the circle outside of the study area the larger will be *k*(*i*,*j*). In the example shown, the boundary is 2.22 miles from point 8 and

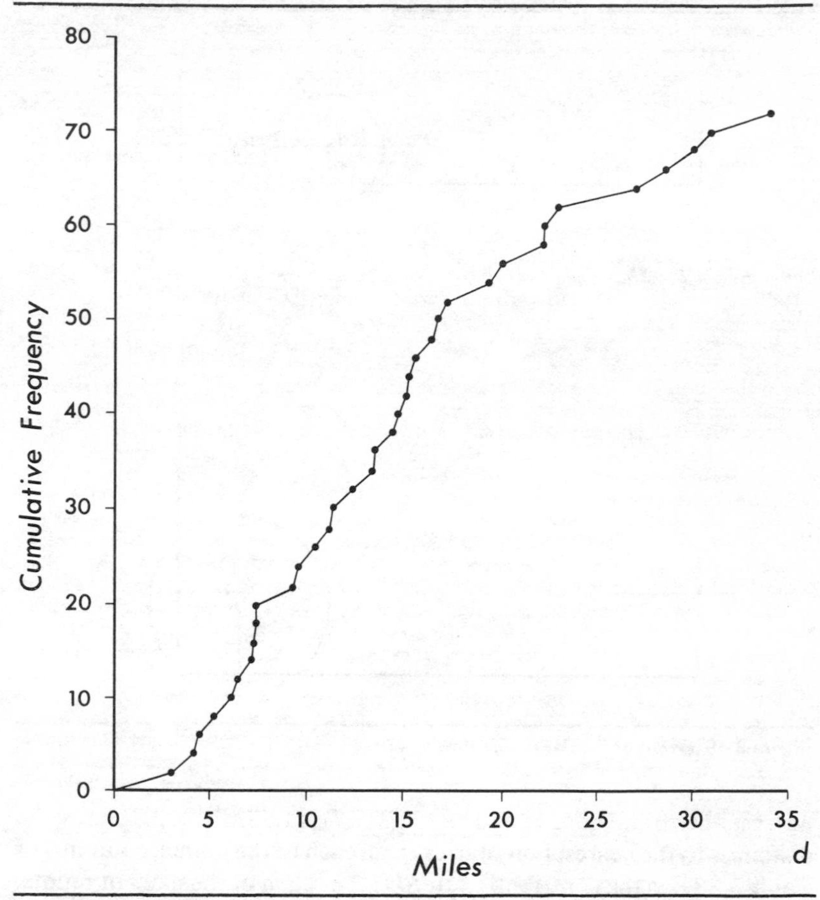

Figure 3.12 Cumulative Frequency of Pairs of Points by Distance for McHenry County

the distance, d, to 7 is 4.12 miles. Thus instead of counting as one pair, $k(8,7)$ counts as 1.47 pairs.

In the case shown in Figure 3.13b, that is, when a point is closer to two right angle boundaries than it is to a point j, the weighting formula given by Getis (1984) is

$$k(i,j) = \{1 - ([\cos^{-1}(e_1/d) + \cos^{-1}(e_2/d) + (\pi/2)]/2\pi)\}^{-1} \qquad [33]$$

In Figure 3.13b, the situation of point 9 is shown in relation to point 7. e_1 is the distance to the nearest vertical border, and e_2 is the distance to the

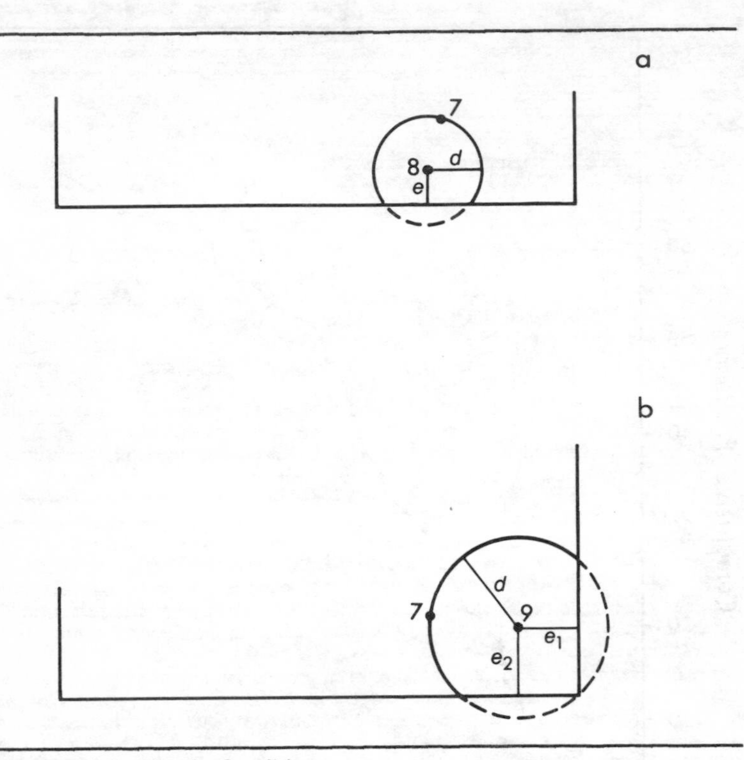

Figure 3.13 Detail of Border Conditions

nearest horizontal border; as a result, $k(9,7)$ is 1.98. Table 3.9 shows the distances to the nearest boundaries from each of the points; column 4 of Table 3.8 gives the contribution to $Sk(i,j)$ of each of the pairs of points.

The two edge correction formulae are inappropriate for irregular or curved borders. One can, however, estimate the value $k(i,j)$ in those circumstances by recalling that $k(i,j)$ is the ratio of the entire circumference to the proportion of the circumference inside the study area. Because of points near corners it might be wise to question the inclusion of values $k(i,j)$ greater than 4. In any case, the border correction device only helps to create unbiasedness for distances d less than the circumradius of the study area or, more conservatively, one-half the lesser of the length or the width of the study area when the study area is a rectangle. For McHenry County, which is 38.06 miles by 33.89 miles, the circumradius is 25.48 miles ($d = \sqrt{((38.06/2)^2 + (33.89/2)^2})$. In the succeeding analysis we will use 25 miles as the cutoff distance.

McHenry County has an area, A, of 1289.81 square miles. When this value is placed into equation 30, $L(d)$ can be calculated for each

TABLE 3.9 Straight-Line Distances in Miles from the Population Centers to the Nearest Boundaries of McHenry County

Population Center	e_1 Nearest Vertical Border	e_2 Nearest Horizontal Border
1	7.78	8.89
2	9.44	11.11
3	16.67	19.44
4	6.39	11.11
5	5.28	10.28
6	10.00	8.89
7	10.56	5.56
8	11.11	2.22
9	4.44	4.44

observed distance (see Table 3.8, column 6). Recall that $L(d)$ represents the expected value in CSR. By comparing the various $L(d)$ with d, one immediately gets an indication of the second-order structure of the interevent distances (see Table 3.8, column 7). An $L(d)$ greater than d at short distances indicates clustering, whereas an $L(d)$ less than d at short distances indicates an inhibition effect, that is, points are consistently separated from one another. At long distances interpretation is more complex because results there are dependent on the short distances outcomes.

Figure 3.14 shows the second-order pattern of population centers in McHenry County. First, an inhibition distance of 3 miles is evident, followed by what appears to be clustering that reaches a peak at 7.4 miles and begins to fade, especially beyond 16 miles. The inhibition results from the obvious fact that population centers represent a spatial cluster of population that precludes the location of another center of population nearby. The peak indicates that there is a tendency for a short-distance spacing regularity of 7.4 miles. Since a small sample was used, this phenomenon is readily evident in the southeast quadrant of the county. The decline in clustering at the longer distances is as much a function of the short distance clustering as it is of the tendency for several of the points to be rather far removed from the more noticeable cluster. In most studies, concern is with short distance $L(d)$. Longer-distance values tend to be blurred by the effect of the short-distance measurements.

The statistical test on the hypothesis of CSR can be carried out by engaging in a rather elaborate series of simulation experiments. The procedure that is usually followed is to use a computer to simulate as many as 99 samples of Poisson process-generated point patterns of size

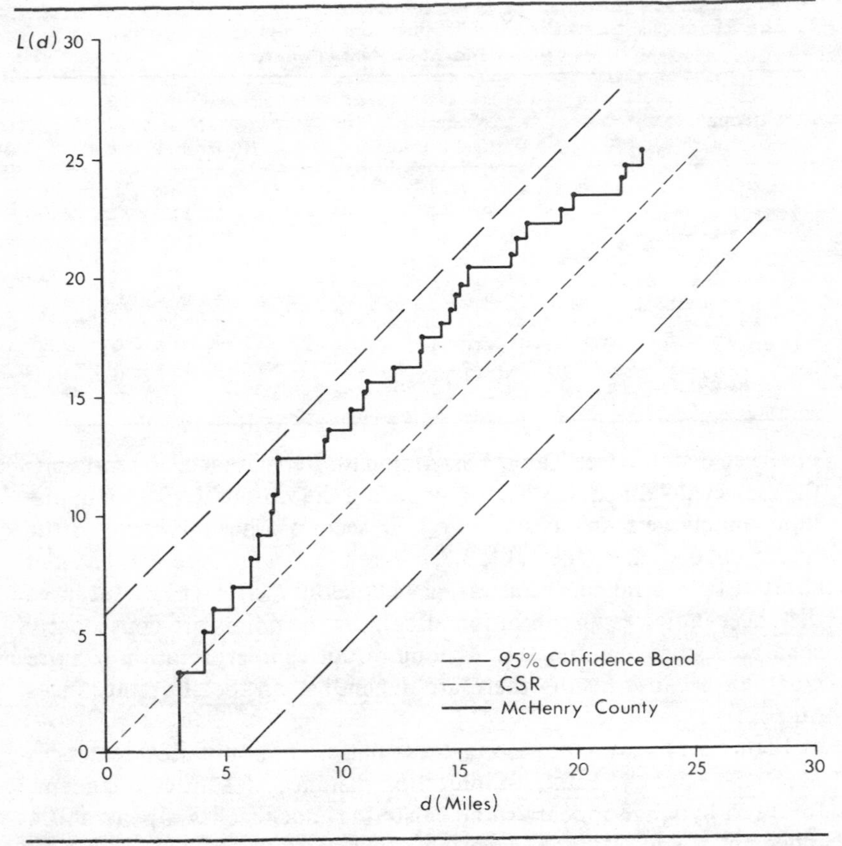

Figure 3.14 Observed and Expected $L(d)$ and the .05 Level of Statistical Significance for Population Centers in McHenry County

N and to compare the extreme case results [the max and min of $L(d)$ of the simulations with the observed values of $L(d)$ for all d]. If the observed $L(d)$ either exceeds the maximum or is less than the minimum for any value of d, the H_0 of CSR is rejected at the .01 level. Of course, this may become a time-consuming and expensive procedure; therefore, we will suggest a much simpler approach that has been developed by Ripley (1979a).

Based on his experience with many simulations, Ripley concluded that a good test at the .05 level is $T = \pm 1.45 A^{1/2}/N$. If any value of $L(d) - d$ is positively or negatively greater than T, statistical significance obtains at the .05 level. Again, high positive values indicate the rejection of a CSR model, with the implication that a model of clustering would be more appropriate. A high negative value indicates the rejection of a

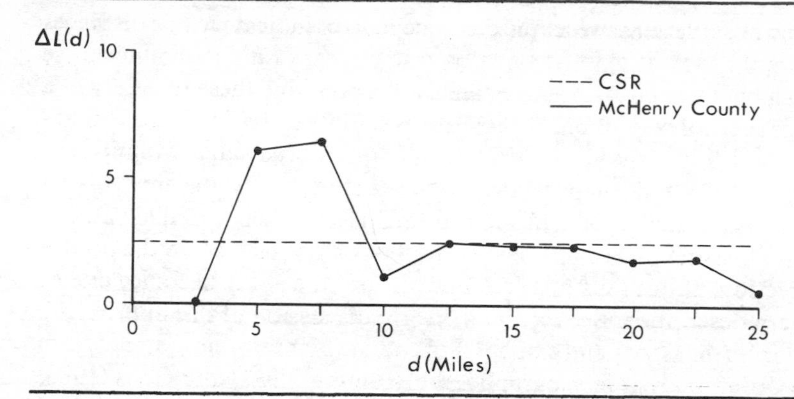

Figure 3.15 Observed and Expected $\Delta L\,(d)$ for Population Centers in McHenry County

CSR model, with the suggestion that a model of evenness or inhibition would be more appropriate. In the McHenry County case the largest difference between $L(d)$ and d is 5.05 when d = 7.43 miles, but the test value is T = 5.79, and thus we conclude that the CSR process cannot be rejected. Most likely, the small sample contributed to this result. The bands representing the .05 level are shown on Figure 3.14.

The remaining aspect of this analysis concerns the study of the number of pairs of points within specified equal-size distance bands. For the McHenry County study, arbitrarily, we have selected zones of 5 miles. By taking the cumulative curve of Figure 3.14 and subdividing it into 10 distance zones, we see on Figure 3.15 the tendency for clustering in each of the zones. We denote $\Delta L(d)$ as the difference in $L(d)$ taking the distance at the two ends of the zone as the cutoffs. Figure 3.15 was constructed by partitioning Table 3.8 into classes, the boundaries of which are d values divisible by 2.5. For example, when d = 5 and 7.5, the values of $L(d)$ are 6.07 and 12.48, respectively. This means that as the expectation rose by 2.5, the observed value rose by (12.48 – 6.07), or 6.41. This is 3.91 above the expectation of 2.5, and is thus an indication of the distinct clustering tendency therein. Inhibition is clearly evident in the zone 0 to 2.5 miles. At the longer distances, the tendency for evenness becomes evident. Of course, one could have chosen smaller zones for more detailed analysis, and for larger N that strategy seems to make more sense. At present, no test exists for the significance of these results, although a difference of proportions test might be indicative of the level of significance.

In an analysis such as that just described, the researcher has to select a distance increment for study. In the McHenry County case, since we had so few observations, we let the distance matrix indicate which distances we would look at closely. Usually, one chooses rather small distance increments to study, since one can always combine these for study at a larger scale of resolution. A suggestion is that the basic distance increment be 1/100 of the length of a side of a square study area. Many times, for simplicity and comparability between study areas, the area A is set equal to one. In that case, all measurements are taken with that value in mind. Another choice one has is whether or not to include the border correction. If one fails to select this option, the study becomes one of pattern description. See section 4.3 for a discussion of this approach.

For further study and examples of the use of interevent distance/second-order analysis in geography, see Haining (1982) and Getis (1983, 1984). The theoretical literature, with many practical examples taken mainly from biology, is thoroughly discussed in Ripley (1981) and Diggle (1983).

4. MEASURES OF ARRANGEMENT

Measures of arrangement are techniques that examine characteristics of the locations of points relative to other points in the pattern. From a practical standpoint, these techniques have two advantages over measures of dispersion discussed in Chapter 3. First, arrangement measures are "density free." This means that in order to obtain arrangement properties of a CSR pattern against which the observed properties are compared, we do not need to estimate any values from the observed data. In the case of quadrat analysis, for example (see section 2.1), it was necessary to estimate λ (the expected number of points per quadrat) before we could obtain the expected quadrat frequencies for a CSR pattern. In addition, since arrangement measures are concerned with the locations of the points relative to each other and not relative to the study area, as is the case with dispersion measures, they are essentially free of edge effects introduced by the nature and shape of the study area boundary.

Measures of arrangement are not without their problems, however. First, such measures are usually less rigorous than measures of dispersion, in something like the way that nonparametric statistics are usually less powerful than their parametric equivalents. This situation arises primarily because measures of arrangement are insensitive to some differences in some pattern characteristics so that identical values may be expected for patterns that are different in some way. In addition, the statistical theory underlying measures of arrangement is generally less well developed than that for measures of dispersion, so that a greater element of subjectivity enters into the interpretation of the results of analyses involving measures of arrangement.

4.1 Reflexive Nearest Neighbor Analysis

In some of the nearest neighbor analyses in Chapter 3 the reader may have observed that there were instances where two points were the nearest neighbors of each other. Such points are said to be reflexive (reciprocal) nearest neighbors. This notion can be extended to identify higher-order reflexive nearest neighbors. One simple test of arrangement involves comparing the number of reflexive nearest neighbors in the pattern under study with that expected in a CSR pattern.

To illustrate this technique, consider Figure 4.1, which shows the locations in a portion of northern Ontario of lakes that in 1978, because of pollution, contained one or more species of fish that were unsafe for human consumption. Of the 57 points in this pattern, 36 have reflexive first nearest neighbors (see Figure 4.1). The probability that a point in a CSR pattern is a reflexive nearest neighbor is 0.6215 (Clark and Evans 1955). Multiplying this probability by N gives the expected number of reflexive nearest neighbors in a CSR pattern, which here is (57) (0.6215) = 35.43. Note that since a reflexive point must be a member of a pair, the observed value must always be an even number. In this example, if we round the expected value to the nearest even number, we get 36, which is identical to the number observed in the empirical pattern. Unfortunately, no test is available to evaluate the significance of the difference between the observed and expected values. Moreover, although a high proportion of reflexive pairs implies that the points occur as isolated and relatively uniformly arranged couples (as in Figure 3.2), it is not entirely clear how to interpret other departures from the CSR expectation (Porter 1960; Pielou 1969: 122; Haggett et al. 1977: 442-445).

Because of both the lack of a test of significance and the lack of unanimity in interpreting the results, it is common to extend the analysis of reflexive nearest neighbors to higher orders (Dacey 1969; Cox 1981). Table 4.1 gives the probabilities that a point in a random pattern is the jth nearest neighbor of its own jth nearest neighbor (for $j \leq 6$). In interpreting the number of observed pairs in relation to CSR values, most researchers suggest that more higher-order values in excess of the CSR expectations indicate a measure of regularity in the arrangement of points, whereas lower empirical values imply elements of grouping in the pattern. In our example the number of second-order reflexive nearest neighbors is 20; the expected number is (57) (0.3291) = 18.76. For the third-order nearest neighbors the observed number of reflexive

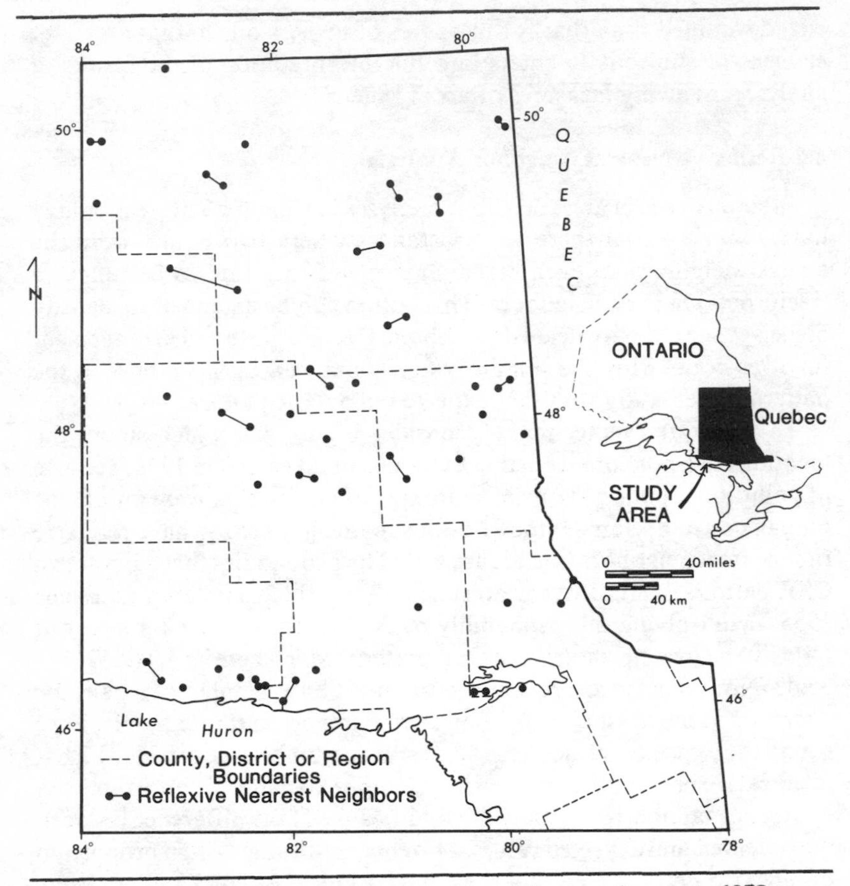

Figure 4.1 Locations of Polluted Lakes in a Part of Northern Ontario in 1978

TABLE 4.1 Probability that a Point in a Random Pattern in Two Dimensions Is the *j*th Nearest Neighbor of Its Own *j*th Nearest Neighbor

j	Probability
1	0.6215
2	0.3291
3	0.2431
4	0.2015
5	0.1760
6	0.1582

SOURCE: Cox (1981).

points is 14 and the expected value is (57) (0.2431) = 13.86. In both cases the difference between the observed and expected values is slight and, together with the values for the first-order reflexive nearest neighbors, suggest that the pattern of polluted lakes is CSR.

The reflexive nearest neighbor procedure may also be applied in those instances where the study area is a line (Dacey 1960). However, this does involve the use of a different set of probabilities for the CSR pattern (Clark 1956). These probabilities are given in Table 4.2.

To illustrate the use of linear reflexive nearest neighbor analysis, consider Figure 4.2, which shows the locations in August 1982 of "fast food" outlets (restaurants providing take-out facilities) on the eastern side of King Street, which is the main street passing through the twin cities of Kitchener-Waterloo in southern Ontario. Of the 32 points, 22 are reflexive nearest neighbors (see Figure 4.2). The expected number for a CSR pattern is (32) (0.6667) = 21.33, thus suggesting that the arrangement characteristics of the pattern are essentially the same as those of a CSR pattern. The observed numbers of second- and third-order reflexive nearest neighbors are 10 and 6, respectively, with the corresponding expected values being 11.84 and 8.70. This suggests that there are some tendencies toward grouping of points in the pattern.

4.2 Polygon Techniques

Polygon techniques are still being developed in geography but the present indications are that they are likely to prove much more sensitive to various pattern arrangements than the reflexive nearest neighbor techniques presented above. However, the polygon techniques do require much more computational effort.

To illustrate these procedures, consider Figure 4.3, which shows the locations in 1973 of settlements in an area of southeastern Montana. For any point pattern in two dimensions, such as this one, it is possible to construct a set of Thiessen polygons, more readily known outside of geography as Voronoi polygons or Dirichlet domains. These polygons are obtained by associating with each point in the pattern all locations in the study area that are closer to it than to any other point in the pattern. Those locations in the study area that are equidistant from two points will lie on the boundary of two adjacent polygons. Similarly, any locations that are equidistant from three or more points in the pattern will form the vertices of adjacent polygons. The result of this procedure is the creation of a tessellation of contiguous, space-exhaustive polygons. The Thiessen polygons for the 30 settlements are shown in Figure 4.3. Examination of these polygons reveals that three edges are incident at each vertex. The incidence of more than three edges at a vertex is very rare since for this to occur four neighboring points in the pattern must lie on the circumference of a circle whose interior contains no other points.

TABLE 4.2 Probability that a Point in a Random Pattern Along a Line is the *j*th Nearest Neighbor of Its Own *j*th Nearest Neighbor

j	Probability
1	0.6667
2	0.3704
3	0.2716
4	0.2241
5	0.1952
6	0.1753

SOURCE: Dacey (1969).

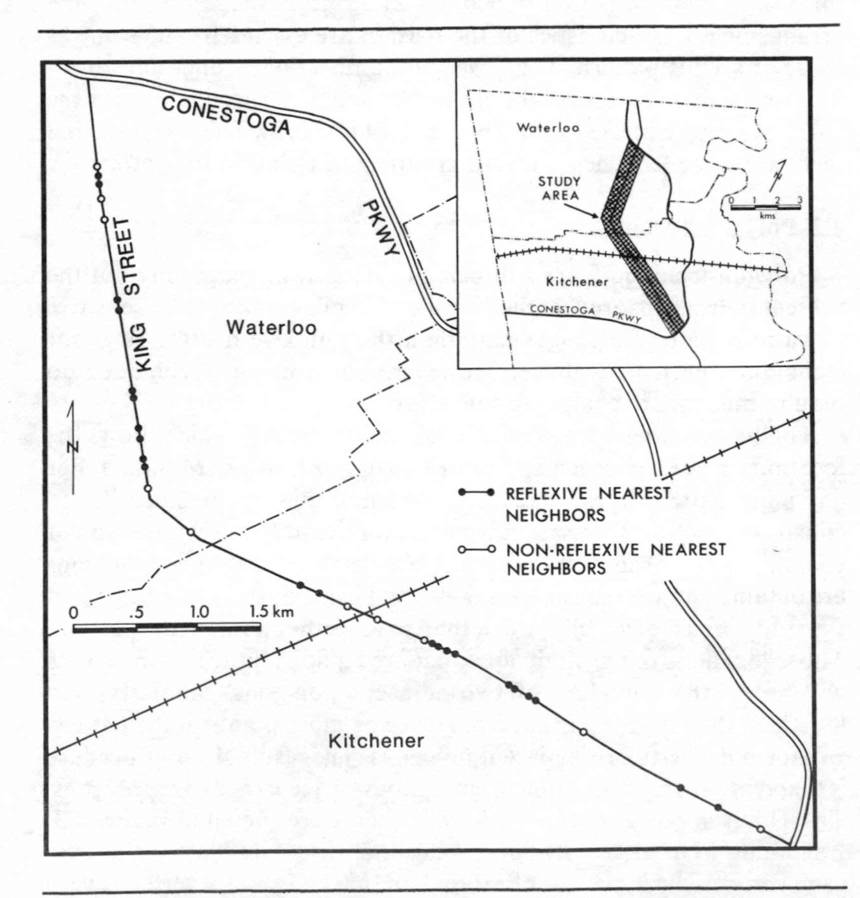

Figure 4.2 Locations of Fast Food Outlets on the Eastern Side of King Street, Kitchener-Waterloo, August 1982

Since this is highly unlikely to occur in real-world patterns, we may discount this situation.

The set of Thiessen polygons can be used to produce another contiguous, space-exhaustive tessellation known as the Delaunay triangulation. This involves joining all points in the pattern whose Thiessen polygons have an edge in common. The resulting tessellation consists exclusively of triangles. The Delaunay triangulation for the Montana settlements is also shown in Figure 4.3.

It is possible to examine several different properties of these triangles but the most obvious density-free property is the sizes of the angles of the triangles. Each triangle has three angles and it is possible to identify the smallest one, which, of course, cannot be greater than 60 degrees. Mardia et al. (1977) have produced results that enable us to determine what the probability is of obtaining a triangle minimum angle less than or equal to some value, x, for a Delaunay triangulation associated with a CSR pattern. This probability, $P(x)$, is given by

$$P(x) = 1 + 1/2\pi\ [(6x - 2\pi) \cos 2x - \sin 2x - \sin 4x] \qquad [34]$$

Values of $P(x)$ for between 1 and 60 degrees (in one-degree intervals) derived from equation 34 are given in Table 4.3.

The specific test involves generating the Delaunay triangulation for the point pattern under investigation and then identifying the minimum angle in each of the triangles in this triangulation. We count the number of minimum angles in the pattern in specified intervals of x. These counts for the triangles in Figure 4.3 are given in column 2 of Table 4.4 Note that in counting the observed frequencies, in order to avoid edge effects, we ignore the angles from those triangles where one or more of the sides forms part of the outer boundary of the Delaunay triangulation of the point pattern. The observed frequencies are then cumulated (see column 3 of Table 4.4) and $F(x)$ is calculated by dividing each of the observed cumulative frequencies by the sum of the frequencies. The values of $F(x)$ so obtained are shown in column 4 of Table 4.4. The corresponding values, $P(x)$, for a CSR pattern can be obtained from column 3 of Table 4.3 and are shown in column 5 of Table 4.4. The values of $F(x)$ and $P(x)$ can be compared using a one-sample Kolmogorov-Smirnov (K-S) test (Yeates 1974: 204-206; Taylor 1977: 137-139; Norcliffe 1977: 102-107; Shaw and Wheeler 1985: 129-130). This test involves obtaining the absolute difference between the values of $F(x)$ and $P(x)$ for corresponding values of x (see column 6 of Table 4.4) The largest of the values in column 6 determines the test statistic, D_{max},

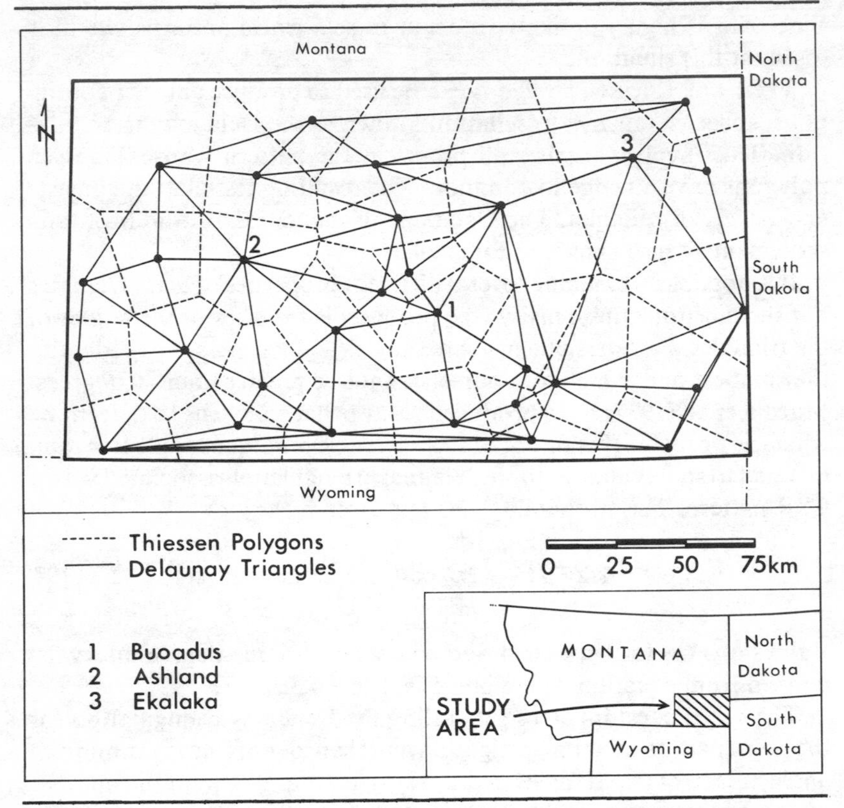

Figure 4.3 Locations of Settlements in Southeastern Montana, 1973

which is compared with the appropriate value from statistical tables of critical values.

For the pattern in Figure 4.3, D_{max} is 0.0680, whereas the critical value from the statistical tables for $\alpha = 0.05$ is 0.2178. Thus we cannot reject the H_0 of a CSR pattern for the settlements.

However, could we have proceeded any farther if the results had led to the rejection of the H_0? The answer is a tentative yes. If the points were arranged perfectly regularly (as in Figure 1.2c) all the Thiessen polygons would be regular hexagons and all the triangles would be equilateral so that all angles in the pattern would measure 60 degrees. Thus, if for a real-world pattern, the H_0 is rejected and there is an excess (relative to the proportions expected for a CSR pattern) of minimum angles at the upper tail, a regular pattern is indicated. Similarly, if the points are located so that they approximate a square grid (see Figure 4.4, which shows such a pattern produced by displacing points originally located on a square lattice), the resulting Thiessen polygons will all be four-sided

TABLE 4.3 Probability that the Minimum Angle of a Trinagle in a Delaunay Triangulation for a Random Point Pattern is Less than or Equal to a Given Value, *x*

[1] Radians x	[2] Degrees	[3] Probability $P(x)$
0.017453	1	0.000610
0.034906	2	0.002436
0.052359	3	0.005479
0.069812	4	0.009731
0.087265	5	0.015189
0.104718	6	0.021843
0.122171	7	0.029683
0.139624	8	0.038698
0.157077	9	0.048871
0.174530	10	0.060185
0.091983	11	0.072619
0.209436	12	0.086152
0.226889	13	0.100757
0.244342	14	0.116406
0.261795	15	0.133068
0.279248	16	0.150707
0.296701	17	0.169287
0.314154	18	0.188768
0.331607	19	0.209107
0.349060	20	0.230256
0.366513	21	0.252170
0.383966	22	0.274794
0.401419	23	0.298074
0.418872	24	0.321954
0.436325	25	0.346374
0.453778	26	0.371271
0.471231	27	0.396582
0.488684	28	0.422240
0.506137	29	0.448177
0.523590	30	0.474322
0.541043	31	0.500604
0.558496	32	0.526949
0.575949	33	0.553283
0.593402	34	0.579531
0.610855	35	0.605617
0.628308	36	0.631463
0.645761	37	0.656994
0.663214	38	0.682133
0.680667	39	0.706804
0.698120	40	0.730930
0.715573	41	0.754437
0.733026	42	0.777252

Continued

TABLE 4.3 Continued

[1] Radians x	[2] Degrees	[3] Probability $P(x)$
0.750479	43	0.799303
0.767932	44	0.820518
0.785385	45	0.840830
0.802838	46	0.860173
0.820291	47	0.878482
0.837744	48	0.895699
0.855197	49	0.911765
0.872650	50	0.926626
0.890103	51	0.940232
0.907556	52	0.952537
0.925009	53	0.963498
0.942462	54	0.973077
0.959915	55	0.981240
0.977368	56	0.987960
0.994821	57	0.993212
1.012274	58	0.996977
1.029727	59	0.999243
1.047179	60	1.000000

and the triangles will be close to right-angled so that the minimum angles will be approximately 45 degrees. Thus a significant excess of angles close to 45 degrees indicates a pattern that approximates a square grid. Note that column 2 of Table 4.4 indicates that the distribution of x is bimodal with one of the modes occurring at 45 degrees, which suggests that the pattern of settlements possesses some elements of a square grid. In a pattern that has clusters of points some Delaunay triangles will have edges that represent links between points on the periphery of different clusters. Such triangles will be obtuse so that their minimum angles will be small. Thus distributions of x that contain a significant excess of small angles are indicative of a clustered arrangement of points. Examples of the use of polygon techniques are found in Boots (1974, 1975), Vincent et al. (1976, 1983), and Vincent and Howarth (1982).

4.3 Interevent Distances Without Boundaries

The discussion in section 3.6 on interevent distances/second-order analysis is designed to enable researchers to test hypotheses on CSR. In that approach if one disregards the need to correct for statistical bias, tests are invalid, but one is still left with a comprehensive description of the arrangement of points. Thus, by eliminating the correction for

TABLE 4.4 Analysis of Locations of Settlements in Southeastern Montana Using the Minimum Angle Technique

[1] α (Degrees)	[2] Observed Frequency	[3] Cumulative Observed Frequency	[4] Cumulative Observed Proportion $F(x)$	[5] Cumulative Expected Proportion $P(x)$	[6] $\lvert [4] - [5] \rvert$ $\lvert F(x) - P(x) \rvert$
0-5	1	1	0.0256	0.0152	0.0104
6-10	4	5	0.1282	0.0602	0.0680
11-15	0	5	0.1282	0.1331	0.0049
16-20	2	7	0.1795	0.2302	0.0508
21-25	6	13	0.3333	0.3464	0.0131
26-30	8	21	0.5385	0.4743	0.0642
31-35	2	23	0.5897	0.6056	0.0159
36-40	4	27	0.6923	0.7309	0.0381
41-45	7	34	0.8718	0.8408	0.0310
46-50	4	38	0.9744	0.9266	0.0478
51-55	1	39	1.0000	0.9812	0.0188
56-60	0	39	1.0000	1.0000	–

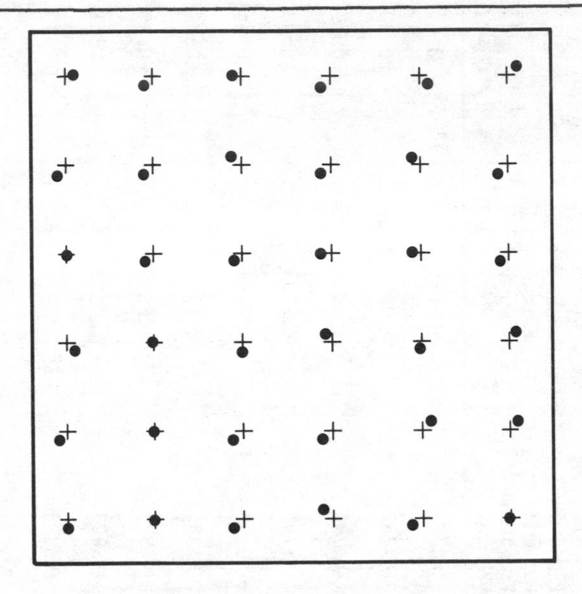

Figure 4.4 A Pattern of Points Approximating a Square Grid

boundaries and dropping concern for CSR—that is, by counting each linked pair as one—we have a measure of spatial arrangement. Rather than confuse this value with $L(d)$, we shall call this measure $A(d)$. The formula for $A(d)$ is

$$A(d) = SI(i,j) \;/\; N(N\text{-}1) \tag{35}$$

where S is defined as in equation 30, $SI(i,j)$ is the sum of all pairs of points within distance d of all i, and N is the total number of points.

The McHenry County data can be used to illustrate the approach outlined in equation 35. In Figure 3.12 the curve of the cumulative frequency of pairs by distance respresents $SI(i,j)$. Figure 4.5 shows $A(d)$. $A(d)$ gives the proportion of all pairs of points within distance d of all i. As was the case with $L(d)$, $A(d)$ may be broken into distance zones [$\Delta A(d)$] so that one may look more carefully at particular trends in the data (Figure 4.6). For example, from Figures 4.5 and 4.6 it can be seen that points are no closer than three miles apart, 20 of the 72 (28%) pairs

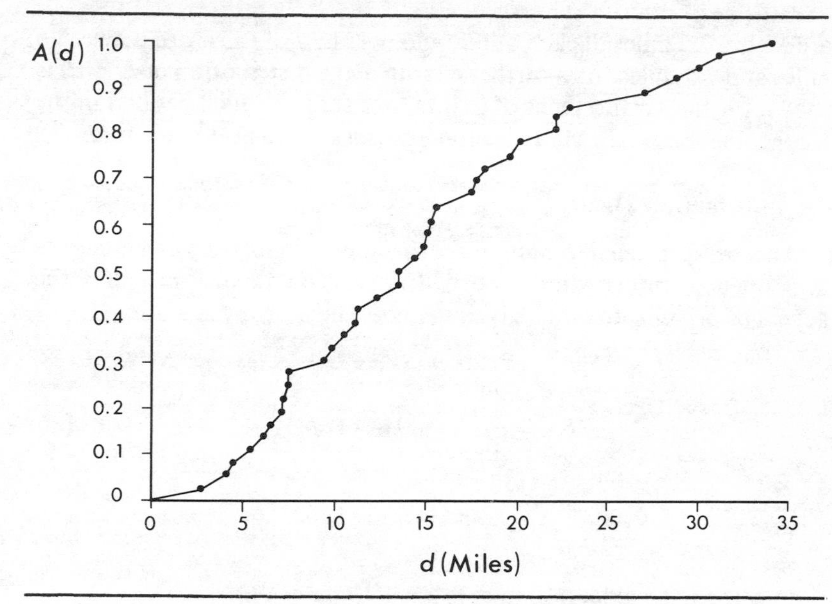

Figure 4.5 *A (d)* **for Population Centers of McHenry County**

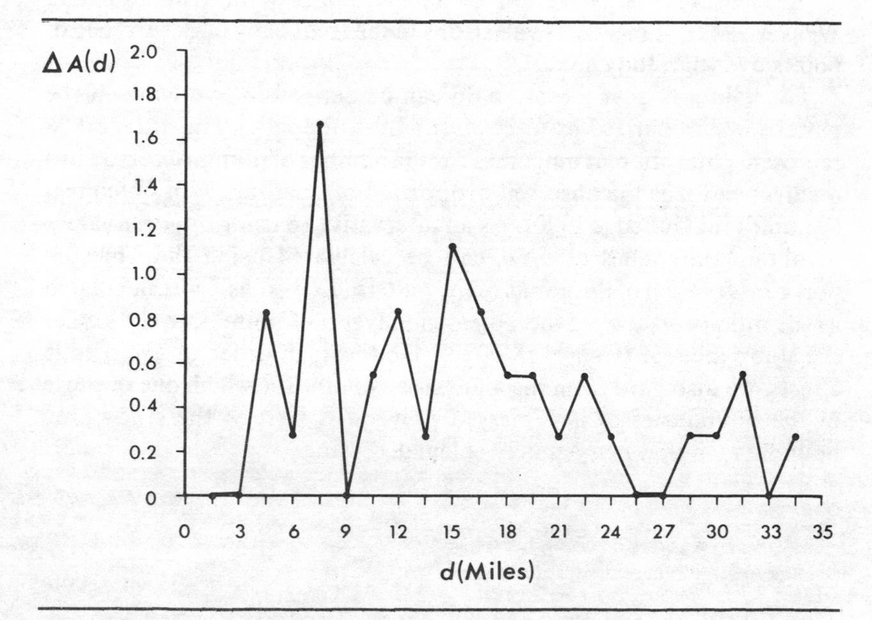

Figure 4.6 *ΔA (d)* **for Population Centers of McHenry County**

are within 7.5 miles of each other, and over half of those are between 6 miles and 7.5 miles. As a further example, the distribution of $\Delta A(d)$ is shown for the arrangement of points (one for each 1000 people) in the Des Moines Standard Metropolitan Statistical Area in 1980 (Figure 4.7).

4.4 Information Theory

Another approach to point pattern analysis involves using concepts developed in information theory. Information theory measures the degree of organization in a given system. The entropy statistic, H

$$H = \sum_{i=1}^{N} p_i \ln (1/p_i) \qquad [36]$$

where:

ln is the natural logarithm and
p_i is defined in the following text

is one measure of organization for a pattern of N points (Shannon and Weaver 1949). It measures variations in the frequency of occurrence of points over the study area.

The values of p_i in equation 36 can be derived in two ways. If the pattern is summarized by a set of quadrats, the index i in equation 36 represents the different outcomes for the number of points observed in a quadrat and p_i is the observed proportion of quadrats with outcome i (Semple and Golledge 1970). As an alternative to using quadrat values to obtain the values of p_i, p_i can be calculated using the Thiessen polygons for each of the points in the pattern. In this case, p_i is calculated as the ratio of size, a_i, of the Thiessen polygon of point, i, to the size of the study area, A (Chapman 1970). However, in order to avoid edge effects, we disregard from the analysis any point for which one or more of the boundaries of its Thiessen polygon is part of the study area boundary, so that p_i is actually calculated using

$$p_i = a_i \Big/ \sum_{i=1}^{n_c} a_i$$

where n_c is the number of points free from edge effects.

There are several problems involved in interpreting the value of H in

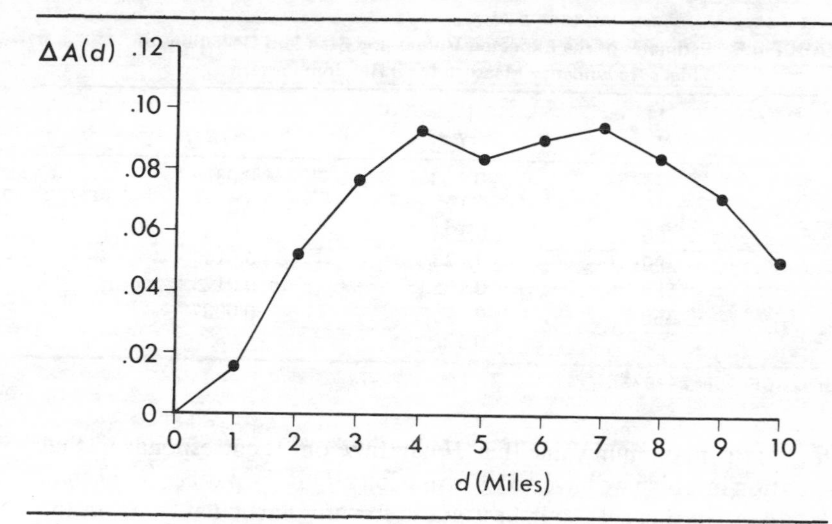

Figure 4.7 $\Delta A(d)$ **for the Population Pattern of Des Moines, Iowa**

equation 36 for a given point pattern. The main problem is that marked clustering and extreme evenness among points in a pattern both yield very similar values of H, whereas CSR patterns with different numbers of points yield drastically different values of H. To overcome these problems, statistics that measure information in a relative way have been proposed.

One such statistic that may be used if the quadrat method is employed to evaluate p_i is the information gain statistic, $I(p{:}q)$. This statistic is given by

$$I(p{:}q) = \sum_i p_i \ln p_i / q_i \quad [38]$$

where q_i is the expected proportion of quadrats with i points in a CSR pattern. Values of q_i can be obtained from equation 1 in section 2.1. The value of two times $I(p{:}q)$ can be evaluated as a χ^2 statistic with $(K - 2)$ degrees of freedom, where K is the number of values of i (Kalbfleisch 1979: 167).

When the Thiessen polygon approach is used, $I(p{:}q)$ cannot be evaluated since q_i has no meaning in this context. In such situations another measure of relative information can be used. This is the redundancy measure, R^*, derived by Theil (1967: 92) and given by

$$R^* = H_{max} - H. \quad [39]$$

TABLE 4.5 Estimates of the Expected Values and Standard Deviations of Thiel's Redundancy Measure for a Random Pattern

[1] N	[2] $E(R^*)$	[3] $Var(R^*)$
15	0.118	0.0389
30	0.122	0.0286
60	0.124	0.0211
90	0.125	0.0165
180	0.125	0.0123
360	0.125	0.0087
∞	0.126	0

SOURCE: Lenz (1979).

H_{max} is the maximum value that H can take on. It corresponds to that situation in which we have no information about the pattern other than N and A. In this case, our best guess for any particular point in the pattern would be that $a_i = A/N$. This would mean that p_i would equal $1/N$ for all N points so that

$$H_{max} = N\{1/N \ \ln\ [1/(1/N)]\} = \ln N \qquad [40]$$

Once computed, the value of R^* is compared to that expected for a CSR pattern, E(R^*). Estimates of E (R^*) and var(R^*) have been obtained by Lenz (1979) and are given in columns 2 and 3, respectively, of Table 4.5. In addition, Lenz suggests that, if $N > 15$, the difference between R^* and E(R^*) can be tested using a normally distributed statistic, z, of the form form

$$z = [R^* - E(R^*)] \ / \ \sqrt{\text{var}\ (R^*)} \qquad [41]$$

Note that $R^* = 0$ for a regular pattern and that it increases as the clustering of points in the study area increases.

A more detailed discussion of the theory underlying these procedures appears in Getis and Boots (1978: 81-85, 142-144) and Lenz (1979); additional empirical examples include Medvedkov (1967, 1970), Semple (1973), Garrison and Paulson (1973), Chapman (1973), and Haynes and Enders (1975).

5. SUMMARY

In this book the reader has been introduced to a variety of techniques for examining point patterns and he or she may now be prompted to ask, "Which one is best?" The answer is that there is no one single optimal technique for all applications and that the selection of a particular procedure is influenced by both practical and statistical concerns. Consequently, we offer in the following text some suggestions that are intended to help the reader in deciding upon an "appropriate" procedure to adopt in any given situation. We stress however, that these suggestions are in the nature of pointers since formal comparative studies are few. In addition, with the exception of those studies by Diggle (1979b) and Ripley (1979b), these comparisons are limited in terms of the extent of techniques and pattern types examined. However, enough is known to indicate that the power of most point pattern techniques (that is, their ability to eliminate false hypotheses) varies according to the type of pattern under examination so that some techniques are better than others in detecting clustering, whereas others are better in detecting regularity.

In general, techniques that measure the dispersion characteristics of a point pattern appear better than those that measure its arrangement properties, particularly since the latter techniques require more subjectivity in the interpretation of their results. Such dispersion measures include the quadrat and distance techniques discussed in Chapters 2 and 3, respectively.

However, when used in association with dispersion techniques, arrangement techniques such as the reciprocal nearest neighbor, polygon, and information theory techniques discussed in Chapter 4 can provide confirmation of results and sometimes also provide additional insight into the pattern. In addition, it is safer to use arrangement techniques when the researcher considers it undesirable to estimate values from the point pattern or when no obvious boundary can be defined for the point pattern.

Within the set of dispersion techniques, quadrat methods require fewer calculations to be performed than distance methods in order to obtain the test statistics. However, if the researcher has access to a digitizer and a computer, this advantage becomes negligible. Further, if the size of N is small, distance techniques are usually more practical. In such circumstances those procedures such as second-order analysis that use all the interevent distances obviously make more use of the limited available information than those procedures that use only average values. Also, as we saw in section 2.2, quadrats suffer from the spatial arrangement problem, so that in many instances it is necessary to perform some additional test of the extent of spatial autocorrelation in the quadrat values. Studies of interevent distance can be rather comprehensive, but one can lose sight of nearest neighbor properties. One might think of second-order and nearest neighbor analysis as complementary, and thus a particularly sensitive analysis might include both.

If the goal of the researcher is simply to evaluate an H_0 of a CSR pattern, distance measures appear more powerful than quadrat analysis (Holgate, 1972). However, quadrat analysis may be preferable if the researcher plans on testing some alternative H_0 since results have been derived for a range of alternative models specified in quadrat form (see Rogers 1974; Haggett et al. 1977: chap. 13; Getis and Boots 1978: chaps. 3 and 4). Few such analytical results have been derived for distance procedures. However, if a computer is available, we may attempt to simulate the process underlying the hypothesized pattern and use a Monte Carlo testing procedure such as that discussed in section 3.5. Recall that this involves measuring one or more properties of the pattern under investigation and considering this pattern to be the outcome of a specified process. This process is then simulated in order to obtain a number of other patterns (usually 99 to correspond with conventional significance levels). The same properties are measured for each of the simulated patterns. We can then examine where the value for the real pattern falls within the entire set of 100 values (99 simulated plus one real pattern) thus giving an indication of the likelihood of the real pattern occurring under the conditions of the specified process. A more detailed discussion of this testing procedure is given by Diggle (1979a).

Finally, we should point out that point pattern analysis is one area in which ongoing research is continuing and indeed expanding, particularly as statisticians become more aware of the area and its challenges. This means that new techniques of point pattern analysis are being developed and existing ones are being refined. Consequently, as we stated in Chapter 1, the particular techniques discussed in this book should be considered as a selection of those presently available.

NOTE

1. Alternatively, once $p(0)$ has been found, successive values of $p(x)$ can be found using the following recursive equation (Pielou 1974: 139):

$$p(x + 1) = [\lambda \,/\, (x + 1)]\, p(x) \quad x = 0, 1, 2, \ldots$$

REFERENCES

Adams, R.E.W., and Jones, R. C. 1981. Spatial patterns and regional growth among Classic Mayan cities. *American Antiquity* 46: 301-322.

Artle, R. K. 1965. *The structure of the Stockholm economy*. Ithaca, NY: Cornell University Press.

Bardsley, W. E. 1978. Note on the distribution of "random" tors on a line. *Mathematical Geology* 10: 375-378.

Birch, B. P. 1967. The measurement of dispersed patterns of settlement. *Tijdschrift voor Economische en Sociale Geografie* 58: 68-75.

Boots, B. N. 1974. Delaunay triangles: an alternative approach to point pattern analysis. *Proceedings, Association of American Geographers* 6: 26-29.

Boots, B. N. 1975. Patterns of urban settlements revisited. *Professional Geographer* 27: 426-431.

Boots, B. N., and Getis, A. 1977. Probability model approach to map pattern analysis. *Progress in Human Geography* 1: 264-286.

Brewer, R., and McCann, M. T. 1985. Spacing in acorn woodpeckers. *Ecology* 66: 307-308.

Burgess, J. W. 1983. Reply to a comment by Mumme et al. *Ecology* 64: 1307-1308.

Burgess, J. W., Roulston, D., and Shaw, E. 1982. Territorial aggregation: an ecological spacing strategy in acorn woodpeckers. *Ecology* 63: 575-578.

Campbell, D. J., and Clark, D. J. 1971. Nearest-neighbor tests of significance for non-randomness in the spatial distribution of singing crickets. (*Teleogryllus commodus* (Walker)). *Animal Behavior* 19: 750-756.

Carmack, R. M., and Weeks, J. M. 1981. The archaeology and ethnohistory of Utalan: a conjunctive approach. *American Antiquity* 46: 323-341.

Chapman, G. P. 1970. The application of information theory to the analysis of population distributions in space. *Economic Geography* 46: 317-331.

Chapman, G. P. 1973. The spatial organization of the population of the United States and England of Wales. *Economic Geography* 49: 325-343.

Clark, P. J. 1956. Grouping in spatial distributions. *Science* 123: 373-374.

Clark, P. J., and Evans, F. C. 1954. Distance to nearest neighbor as a measure of spatial relationships in populations. *Ecology* 35: 445-453.

Clark, P. J., and Evans, F. C. 1955. On some aspects of spatial patterns in biological populations. *Science* 121: 397-398.

Clark, W.A.V. 1969. Applications of spacing models in intra-city studies. *Geographical Analysis* 1: 391-399.

Cliff, A. D., and Ord, J. K. 1975. Model building and the analysis of spatial pattern in human geography. *Journal of the Royal Statistical Society* Series B 37: 297-348.

Cliff, A. D., and Ord, J. K. 1981. *Spatial processes: models and applications*. London: Pion.

Cormack, R. M. 1979. Spatial aspects of competition between individuals. In *Spatial and temporal analysis in ecology*, eds. R. M. Cormack and J. K. Ord, pp. 151-212. Fairland, MD: International Co-operative Publishing House.

Cowie, S. R. 1967. *The cumulative frequency nearest-neighbour method for the identification of spatial patterns.* Seminar paper, series A, 10, University of Bristol, Department of Geography.

Cox, T. F. 1981. Reflexive nearest neighbors. *Biometrics* 37: 367-69.

Dacey, M. F. 1960. The spacing of river towns. *Annals of the Association of American Geographers* 50: 59-61.

Dacey, M. F. 1962. Analysis of central places and point pattern by a nearest neighbor method. *Lund Studies in Geography* 24: 55-75.

Dacey, M. F. 1963. Order neighbor statistics for a class of random patterns in multidimensional space. *Annals of the Association of American Geographers* 53: 505-515.

Dacey, M. F. 1964a. Two-dimensional random point patterns: a review and an interpretation. *Papers and Proceedings of the Regional Science Association* 13: 41-55.

Dacey, M. F. 1964b. Modified Poisson probability law for point pattern more regular than random. *Annals of the Association of American Geographers* 54: 559-565.

Dacey, M. F. 1965. Order distance in an inhomogeneous random point pattern. *Canadian Geographer* 9: 144-153.

Dacey, M. F. 1966a. A county seat model for the areal pattern of an urban system. *Geographical Review* 56: 527-542.

Dacey, M. F. 1966b. A compound probability law for a pattern more dispersed than random and with areal inhomogeneity. *Economic Geography* 42: 172-179.

Dacey, M. F. 1966c. A probability model for central place location. *Annals of the Association of American Geographers* 56: 550-568.

Dacey, M. F. 1968. An empirical study of the areal distribution of houses in Puerto Rico. *Transactions of the Institute of British Geographers* 45: 15-30.

Dacey, M. F. 1969a. Proportion of reflexive nth order neighbors in spatial distributions. *Geographical Analysis* 1: 385-388.

Dacey, M. F. 1969b. A hypergeometric family of discrete probability distributions: properties and applications to location models. *Geographical Analysis* 1: 283-317.

Dacey, M. F. 1969c. Some properties of a cluster point process. *Canadian Geographer* 13: 128-140.

Dacey, M. F. 1969d. Some spacing measures of areal point distributions having the circular normal form. *Geographical Analysis* 1: 15-30.

Dacey, M. F. 1969e. Similarities in the areal distributions of houses in Japan and Puerto Rico. *Area* 3: 35-37.

Dacey, M. F. 1973a. A central focus cluster process for urban dispersion. *Journal of Regional Science* 13: 77-90.

Dacey, M. F. 1973b. A specification of stochastic location processes. *Northeast Regional Science Review* 3: 10-16.

Dacey, M. F. 1975. Evaluation of the Poisson approximation to measures of the random pattern in a square. *Geographical Analysis* 7: 351-367.

Dacey, M. F., and Krumbein, W. C. 1971. *Comments on spatial randomness in dendritic stream channel networks.* Technical Report 17, contract N00014-67-A0356-0018, Department of Geological Sciences, Northwestern University.

Dacey, M. F., and Tung, T.-H. 1962. The identification of randomness in point patterns. *Journal of Regional Science* 4: 83-96.

Day, M. J. 1976. The morphology and hydrology of some Jamaican karst depressions. *Earth Surface Processes* 1: 111-129.

Diggle, P. J. 1976. Note on the Clark and Evans test of spatial randomness. In *Spatial analysis in archaeology,* eds. I. Hodder and C. Orton, pp. 246-248. New York: Cambridge University Press.

Diggle, P. J. 1979a. Statistical methods for spatial point patterns in ecology. In *Spatial and temporal analysis in ecology,* eds. R. M. Cormack and J. K. Ord, pp. 95-150. Fairland, MD: International Co-operative Publishing House.

Diggle, P. J. 1979b. On parameter estimation and goodness-of-fit testing for spatial point patterns. *Biometrics* 35: 87-101.

Diggle, P. J. 1981. Statistical analysis of spatial point patterns. *New Zealand Statistician* 16: 22-41.

Diggle, P. J. 1983. *Statistical analysis of spatial point patterns.* New York: Academic Press.

Donnelly, K. P. 1978. Simulations to determine the variance and edge effect of total nearest neighbour distance. In *Simulation methods in archaeology,* ed. I. Hodder, pp. 91-95. Cambridge: Cambridge University Press.

Durbin, J. 1965. Discussion on Pyke (1965). *Journal of the Royal Statistical Society,* Series B 27: 437-438.

Faniran, A. 1974. Nearest-neighbour analysis of inter inselberg distance: a case study of the inselbergs of south-western Nigeria. *Zeitschrift fur Geomorphologie,* Supplementband 20: 150-167.

Garrison, C. B., and Paulson, A. S. 1973. An entropy measure of the geographic concentration of economic activity. *Economic Geography* 49: 319-324.

Getis, A. 1964. Temporal land-use pattern analysis with the use of nearest neighbor and quadrat methods. *Annals of the Association of American Geographers* 54: 391-399.

Getis, A. 1974. Representation of spatial point processes by Polya models. In *Proceedings of the 1972 meeting of the IGU Commission on Quantitative Geography,* ed. M.H. Yeates, pp. 76-100. Montreal: McGill-Queen's University Press.

Getis, A. 1983. Second-order analysis of point patterns: the case of Chicago as a multi-center urban region. *Professional Geographer* 35: 73-80.

Getis, A. 1984. Interaction modeling using second-order analysis. *Environment and Planning* A 16: 173-183.

Getis, A., and Boots, B. 1978. *Models of spatial processes: an approach to the study of point, line and area patterns.* Cambridge: Cambridge University Press.

Gravenor, C. P. 1974. The Yarmouth drumlin field, Nova Scotia, Canada. *Journal of Glaciology* 13: 45-54.

Green, F.H.W. 1955. *Local accessibility.* Chessington: Surrey, Director General of the Ordnance Survey.

Greig-Smith, P. 1964. *Quantitative plant ecology.* London: Butterworths.

Griffith, D., and Amrhein, C. 1983. An evaluation of correction techniques for boundary effects in spatial statistical analysis: traditional methods. *Geographical Analysis* 15: 352-360.

Guy, C. 1976. The location of shops in the Reading area. *Reading Geographical Papers* 46.

Haining, R. 1982. Describing and modeling rural settlement maps. *Annals of the Association of American Geographers* 72: 211-223.

Haggett, P., Cliff, A. D., and Frey, A. 1977. *Locational analysis in human geography.* London: Edward Arnold.

Harrison, J. S., and Gentry, J. B. 1981. Foraging pattern, colony distribution, and foraging range of the Florida harvester ant, *Pogonomyrmex badius. Ecology* 62: 1467-1473.

Harvey, D. W. 1966. Geographical processes and the analyses of point patterns. *Transactions of the Institute of British Geographers* 40: 81-95.

Harvey, D. W. 1968a. Some methodological problems in the use of the Neyman Type A and the negative binomial probability distributions for the analysis of spatial point patterns. *Transactions of the Institute of British Geographers* 44: 85-95.

Harvey, D. W. 1968b. Pattern, process and the scale problem in geographical research. *Transactions of the Institute of British Geographers* 45: 71-78.

Haynes, K. E., and Enders, W. T. 1975. Distance, direction, and entropy in the evolution of a settlement pattern. *Economic Geography* 51: 357-365.

Hill, A. R. 1973. The distribution of drumlins in County Down, Ireland. *Annals of the Association of American Geographers* 63: 226-240.

Hodder, I., and Orton, C. 1976. *Spatial analysis in archaeology.* Cambridge: Cambridge University Press.

Holgate, P. 1972. The use of distance methods for the analysis of spatial distributions of points. In *Stochastic point processes,* ed. P.A.W. Lewis, pp. 122-135. New York: John Wiley.

Hsu, S.-Y., and Mason, J. D. 1974. The nearest neighbor statistics for testing randomness of point distributions in a bounded two-dimensional space. In *Proceedings of the 1972 Meeting of the IGU Commission on Quantitative Geography,* ed. M. H. Yeates, pp. 32-54. Montreal: McGill-Queen's University Press.

Hudson, J. C. 1969. Pattern recognition in empirical map analysis. *Journal of Regional Science* 9: 189-199.

Hudson, J. C. 1971. Structure of point processes. *Geographical Analysis* 3: 262-267.

Ingram, D. R. 1978. An evaluation of procedures utilised in nearest-neighbour analysis. *Geografiska Annaler* 60B: 65-70.

Jauhiainen, E. 1975. Morphometric analysis of drumlin fields in northern Central Europe. *Boreas* 4: 219-230.

Jones, A. 1971. An order neighbor approach to random disturbances on regular point lattices. *Geographical Analysis* 3: 361-378.

Kalbfleisch, J. G. 1979. *Probability and statistical inference II.* New York: Springer-Verlag.

King, C.A.M. 1974. Morphometry in glacial geomorphology. In *Glacial Geomorphology*, ed. D. R. Coates, pp. 147-162. Binghamton, NY: State University of New York Press.

King, L. J. 1961. A multivariate analysis of the spacing of urban settlements in the United States. *Annals of the Association of American Geographers* 51: 222-233.

King, L. J. 1962. A quantitative expression of the patterns of urban settlement in selected areas of the United States. *Tijdschrift voor Economische en Sociale Geografie* 53: 1-7.

King, L. J. 1985. *Central place theory.* Newbury Park, CA: Sage.

Lamberti, G. A., and Resh, V. H. 1983. Stream periphyton and insect herbivores: an experimental study of grazing by a caddisfly population. *Ecology* 64: 1124-1135.

Lee, Y. 1974. An analysis of spatial mobility of urban activities in downtown Denver. *Annals of Regional Science* 8: 95-108.

Lenz, R. 1979. Redundancy as an index of change in point pattern analysis. *Geographical Analysis* 11: 374-88.

Levings, S. C. and Franks, N. R. 1982. Patterns of nest dispersion in a tropical ground ant community. *Ecology* 64: 338-344.

Lewis, P.A.W. 1961. Distribution of the Anderson-Darling statistic. *Annals of Mathematical Statistics* 32: 1118-1124.

Malm, R., Olsson, G., and Warneryd, O. 1966. Approaches to simulations of urban growth. *Geografiska Annaler* 48B: 9-22.

Mardia, K. V., Edwards, R., and Puri, M. L. 1977. Analysis of central place theory. *Bulletin of the International Statistical Institute* 47, 2: 93-110.

Matui, I. 1932. Statistical study of the distribution of scattered villages in two regions of the Tonami Plain, Toyama Prefecture. *Japanese Journal of Geology and Geography* 9: 251-266. (Reprinted in *Spatial analysis: a reader in statistical geography,* eds. B.J.L. Berry and D. F. Marble, pp. 149-158. Englewood Cliffs, NJ: Prentice-Hall.)

McClure, M. S. 1976. Spatial distribution of pit-making ant lion larvae *(Neuroptera myrmeleontodae):* density effects. *Biotropica* 8: 179-183.

McConnell, H., and Horn, J. M. 1972. Probabilities of surface karst. In *Spatial analysis in geomorphology,* ed. R. J. Chorley, pp. 111-133. London: Methuen.

McNutt, C. H. 1981. Nearest neighbors, boundary effect, and the old flag trick: a general solution. *American Antiquity* 46: 571-592.

Medvedkov, Y. V. 1967. Concept of entropy in settlement pattern analysis. *Papers and Proceedings of the Regional Science Association,* 18: 165-168.

Medvedkov, Y. V. 1970. Entropy: an assessment of potentialities in geography. *Economic Geography* 46: 306-316.

Moran, P.A.P. 1950. Notes on continuous stochastic phenomena. *Biometrika* 37: 17-23.

Morgan, R.P.C. 1970. An analysis of basin asymmetry in the Klang basin, Selangor. *Bulletin of the Geological Society of Malaysia* 3: 17-26.

Muller, E. H. 1974. Origins of drumlins. *In Glacial Geomorphology*, ed. D.R. Coates, pp. 187-204. Binghamton, NY: State University of New York.

Mumme, R. L., Koenig, W. D., and Pitelka, F. A. 1983. Are Acorn Woodpecker territories aggregated? *Ecology* 64: 1305-1307.

Neyman, J., and Scott, E. L. 1952. A theory of the spatial distribution of the galaxies. *Astrophysics Journal* 116: 144-163.

Neyman, J. and Scott, E. L. 1958. A statistical approach to problems of cosmology. *Journal of the Royal Statistical Society Series* B 20: 1-43.

Neyman, J., Scott, E. L., and Shane, C. A. 1956. Statistics of images of galaxies with particular reference to clustering. *Proceedings of the Third Berkeley Symposium of Mathematical Statistics and Probability* 3: 75-111.

Norcliffe, G. B. 1977. *Inferential statistics for geographers: an introduction.* London: Hutchinson.

Oeppen, B. J. and Ongley, E. D. 1975. Spatial point processes applied to the distribution of river junctions. *Geographical Analysis* 7: 153-171.

Odland, J. 1987. *Spatial autocorrelation.* Newbury Park, CA: Sage.

Olsson, G. 1966. Central place systems, spatial interaction, and stochastic processes. *Papers and Proceedings of the Regional Science Association* 18: 13-45.

Peebles, P.J.E. 1974. The nature of the distribution of galaxies. *Astronomy and Astrophysics* 32: 197-202.

Pielou, E. C. 1969. *An introduction to mathematical ecology.* New York: John Wiley.

Pielou, E. C. 1974. *Population and community ecology: principles and methods.* New York: Gordon and Breach.

Pinder, D., Shimada, I., and Gregory, D. 1979. The nearest-neighbor statistic: archaeological application and new developments. *American Antiquity* 44: 430-445.

Porter, P. W. 1960. Earnest and the Orephagians—a fable for the instruction of young geographers. *Annals of the Association of American Geographers* 50: 297-299.

Ripley, B. D. 1979a. The analysis of geographical maps. In *Exploratory and explanatory statistical analysis of spatial data,* eds. C.P.A. Bartels and R.H. Ketellapper, pp. 53-72. Boston: Nijhoff.

Ripley, B. D. 1979b. Tests of "randomness" for spatial point patterns. *Journal of the Royal Statistical Society,* Series B: 368-374.

Ripley, B. D. 1981. *Spatial statistics.* New York: John Wiley.

Roberts, R. C. 1979. Habitat and resource relationships in Acorn Woodpeckers. *Condor* 81: 1-8.
Robinson, G., Peterson, J. A., and Anderson, P. A. 1971. Trend surface analysis of corrie altitudes in Scotland. *Scottish Geographical Magazine* 87: 142-146.
Roder, W. 1974. Application of a procedure for statistical assessment of points on a line. *Professional Geographer* 26: 283-290.
Roder, W. 1975. A procedure for assessing point patterns without reference to area or density. *Professional Geographer* 27: 432-440.
Rogers, A. 1965. A stochastic analysis of the spatial clustering of retail establishments. *Journal of the American Statistical Association* 60: 1094-1103.
Rogers, A. 1969a. Quadrat analysis of urban dispersion: 1. Theoretical techniques. *Environment and Planning* 1: 47-80.
Rogers, A. 1969b. Quadrat analysis of urban dispersion: 2. Case studies of urban retail systems. *Environment and Planning* 1: 155-171.
Rogers, A. 1974. *Statistical analysis of spatial dispersion: the quadrat method.* London: Pion Press.
Rose, J., and Letzer, J. M. 1975. Drumlin measurements: a test of the reliability of data derived from 1:25000 scale topographic maps. *Geological Magazine* 112: 361-371.
Selkirk, K. E., and Neave, H. R. 1984. Nearest-neighbor analysis of one-dimensional distributions of poins. *Tijdschrift voor Economische en Sociale Geografie* 75: 356-362.
Semple, R. K. 1973. Recent trends in the spatial concentration of corporate headquarters. *Economic Geography* 49: 309-318.
Semple, R. K., and Golledge, R. G. 1970. An analysis of entropy changes in a settlement pattern over time. *Economic Geography* 46: 157-160.
Shannon, C. E., and Weaver, W. 1949. *The mathematical theory of communication.* Urbana: University of Illinois Press.
Shaw, B. 1978. *Processes and patterns in the geography of retail change.* University of Hull, Occasional Papers in Geography, 24.
Shaw, G., and Wheeler, D. 1985. *Statistical techniques in geographical analysis.* Chichester: John Wiley.
Sherwood, K. B. 1970. Some applications of nearest-neighbour technique to the study of intra-urban functions. *Tijdschrift voor Economische en Sociale Geografie* 61: 41-48.
Sibley, D. 1972. Strategy and tactics in the selection of shop locations. *Area* 4: 151-156.
Sibley, D. 1975. *The small shop in the city.* University of Hull, Occasional Papers in Geography 22.
Simberloff, D. S. 1979. Nearest neighbor assessments of spatial configurations of circles rather than points. *Ecology* 60: 679-685.
Simberloff, D. S., Dillon, P., King, L., Lorence, D., Lowrie, S., and Schilling, E. 1978. Holes in the doughnut theory: the dispersion of ant-lions. *Brenesia* 14/15: 13-46.
Smalley, I. J., and Unwin, D. J. 1968. The formation and shape of drumlins and their distributions and orientation in drumlin fields. *Journal of Glaciology* 7: 377-390.
Sneyd, A. D. 1982. An inter-tor distance distribution. *Mathematical Geology* 14: 669-673.
Southwood, T.R.E. 1966. *Ecological methods.* London: Methuen.
Southwood, T.R.E. 1978. *Ecological methods.* New York: Halstead.
Stark, B. L., and Young, D. L. 1981. Linear nearest neighbor analysis. *American Antiquity* 46: 284-300.
Sugden, D. E. 1969. The age and form of corries in the Cairngorms. *Scottish Geographical Magazine* 85: 34-46.
Taylor, P. J. 1977. *Quantitative methods in geography: an introduction to spatial analysis.* Boston: Houghton Mifflin.

Theil, H. 1967. *Economics and information theory.* Amsterdam: North Holland.

Thomas, R. W. 1979. *An introduction to quadrat analysis.* Norwich, England: Geo-abstracts.

Thomas, R. W. 1981. Point pattern analysis. In *Quantitative Geography,* eds. N. Wrigley and R. J. Bennett, pp. 166-176. Henley-on-Thames: Routledge & Kegan Paul.

Thomspon, H. R. 1956. Distribution of distance to nth neighbor in a population of randomly distributed individuals. *Ecology* 37: 391-394.

Tinkler, K. J. 1971. Statistical analysis of tectonic patterns in areal volcanism: the Bunyaruguru volcanic field in west Uganda. *Mathematical Geology* 3: 335-355.

Trenhaile, A. S. 1971. Drumlins: their distribution, orientation and morphology. *Canadian Geographer* 15: 113-126.

Trenhaile, A. S. 1975. The morphology of drumlin fields. *Annals of the Association of American Geographers* 65: 297-312.

Unwin, D. J. 1973. The distribution and orientation of corries in northern Snowdonia, Wales. *Transactions of the Institute of British Geographers* 58: 85-97.

Upton, G., and Fingleton, B. 1985. *Spatial data analysis by example. Volume 1: Point pattern and quantitative data.* Chichester: John Wiley.

Vincent, P., Collins, R., Griffiths, J., and Howarth, J. 1983. Statistical geometry of geographical point patterns. *Geographia Polonica* 45: 109-125.

Vincent, P. J., and Howarth, J. 1982. A theoretical note on spatial pattern of discs and properties of simplicial graph. *Forest Science* 28: 181-186.

Vincent, P. J., Howarth, J. M., Griffiths, J. C., and Collins, R. 1976. The detection of randomness in plant patterns. *Journal of Biogeography* 3: 373-380.

Ward, C. F. 1983. The use of locational analysis at Chisalin, Guatemala. *American Antiquity* 48: 132-135.

Weeks, J. M. 1983. Locational analysis and the perception of social groups in Chisalin, Guatemala. *American Antiquity* 48: 136-138.

Werlefors, T., Eskilsson, C., and Ekelund, S. 1979. A method for the automatic assessment of carbides in high speed steels with a computer controlled scanning electron microscope. *Scandinavian Journal of Metallurgy* 8: 221-231.

White, J., Lloyd, M., and Zar, J. H. 1979. Faulty eclosion in crowded suburban periodical cicadas: populations out of control. *Ecology* 60: 305-315.

Williams, P. W. 1972a. The analysis of spatial characteristics of karst terrains. In *Spatial Analysis of Geomorphology,* ed. R. J. Chorley, pp. 135-165. New York: Harper & Row.

Williams, P. W. 1972b. Morphometric analysis of polygonal karst in New Guinea. *Bulletin of the Geological Society of America* 83: 761-796.

Wray, P. J., Richmond, O., and Morrison, H. L. 1983. Use of the Dirichlet tessellation for characterizing and modeling nonregular dispersions of second-phase particles. *Metallography* 16: 39-58.

Yeates, M. 1974. *An introduction to quantitative analysis in human geography.* New York: McGraw-Hill.

ABOUT THE AUTHORS

BARRY N. BOOTS is Professor and former Chairman of Geography at Wilfrid Laurier University in Ontario, Canada. He received his Ph.D. in geography from Rutgers University in 1972. His research has focused upon the examination of spatial structure and the foundations for the analysis of patterns of points. A sample of publication outlets for his research reflects not only his eclectic style but the wide range of literature that benefits from his contributions to the analysis of spatial patterns: *Geographical Analysis, Journal of Glaciology, Environment and Planning A, Metallography,* and *Economic Geography.*

ARTHUR GETIS is Professor and former Head of Geography at the University of Illinois at Urbana-Champaign. He received his Ph.D. in geography from the University of Washington in 1961. He has taught at Rutgers, Princeton, UCLA, and Bristol (United Kingdom). His research has focused upon studies in urban and economic geography, especially that which translates into an analysis of point pattern analysis. He has published widely in a variety of outlets including *Environment and Planning A, Transactions of the Institute of British Geographers, Geographical Analysis,* and *Urban Geography*. He has written several geography textbooks; his interest in geographic education is reflected also in his writing for educational geography journals and his participation on the editorial board of National Geographic Society's *Geographic Research.* Professors Boots and Getis have collaborated on past research including their coauthorship of *Models of Spatial Processes.* Both authors contributed equally to the preparation of this manuscript.

NOTES